THE ECONOMICS OF THE OIL CRISIS

The Economics of the Oil Crisis

Theories of Oil Crisis, Oil Rent, and Internationalization of Capital in the Oil Industry

by
Cyrus Bina
Professor of Economics, Olivet College, Michigan

ST MARTIN'S PRESS NEW YORK

St Martin's Press, Inc., 175 Fifth Avenue, New York, NY 10010

Printed in Great Britain
First published in the United States of America in 1985

ISBN 0-312-23661-1

Library of Congress Cataloging in Publication Data

Bina, Cyrus
The economics of the oil crisis
Bibliography: P.
Includes index
1. Petroleum industry and trade. I. Title
HD 9560.5B47 1985 338.2'7282 84-22888

ISBN 0-312-23661-1

'Capitalist production constantly strives to overcome the immanent barriers, but it overcomes them only by means that set up these barriers afresh and on a more powerful scale. The *true barrier* to capitalist production is *capital itself*.'

Karl Marx, *CAPITAL*, Volume 3 (New York, Vintage Books, 1981), p. 358.

'It's easy to accuse the Arab nations of being "greedy" because they are trying to get what the market will bear for their oil. But it is well to bear in mind that where self-interest is concerned, all of us are brothers under the skin.'

FORBES MAGAZINE, June 15, 1973, p. 88.

TABLE OF CONTENTS

PART TWO

THEORETICAL DEVELOPMENT AND EMPIRICAL ANALYSIS OF THE OIL CRISIS

ACKNOWLEDGEMENTS

The original idea of conducting the present study is perhaps as old as the oil crisis itself. Throughout these intervening years I was quite fortunate in being able to debate some of the issues raised in this volume on a number of occasions. It was during these encounters, as well as during the preparation of the manuscript, that I incurred so many debts.

I am particularly indebted to Ben Fine of Birkbeck College, University of London, to Willi Semmler of New School for Social Research, and most of all to Anwar Shaikh of New School for Social Research for their extensive comments, constructive criticism, and personal encouragement. My special thanks are due to John Willoughby of the American University for his painstaking reading of the entire manuscript and invaluable suggestions. I also wish to thank both Robin Hahnel and John Weeks of the American University for their helpful comments on the earlier drafts of the manuscript. I also owe special thanks to Walter Measday of the U.S. Senate Subcommittee on Antitrust and Monopoly for his comments on the empirical sections and sharing of his invaluable experience with me on a number of occasions.

I wish to express my appreciation to the staff members of the U.S. Department of Energy Library, the U.S. Department of the Interior Library, the American Petroleum Institute Library, and the Library of Congress. My deep appreciation is extended to Ozzie Simms of the U.S. Energy Information Administration for his invaluable assistance and dedication to public service.

Mohammad Kharabi deserves special thanks for his assistance in the usage of computer facilities at the American University and also for his personal encouragement. I am also grateful to Fran Christiansen and Linda Manke for their excellent typing and retyping of the manuscript with a great deal of personal care.

Elisabeth Selden of Olivet College deserves special praise for her editing of the primary drafts of the manuscript. I am also profoundly grateful to Keith Thomas of Olivet College for his personal support and enthusiasm in the final preparation of this book. He also deserves sincere thanks for his assistance in computation of the econometric models. I am quite fortunate to benefit from the poetic touch of Linda J. Scott of Olivet College for her assistance in editing of the final proofs of this study.

I owe a quite specific debt to my wife and children for their patience and tolerance during the entire period of my engagement in preparation of this volume.

Last but not least, this page of dedication honors my mother, a great woman of courage, integrity and sacrifice, whose devotion to her children cannot be expressed in words.

Washington, D.C.
August 1983

Cyrus Bina

LIST OF TABLES

LIST OF ILLUSTRATIONS

PREFACE

The primary focus of this book is to analyze the cause of the world oil crisis of 1973-74. However, the theories and conceptual arguments developed in conjunction with this topic have a broader implication for understanding the laws of motion of oil production, that goes beyond this particular subject.

There are four basic theoretical formulations that will jointly provide the foundation of the alternative theory of the oil crisis developed in this book. These are: (1) theory of oil property relations in the oil industry, (2) theory of internationalization of capital in the oil industry, and historical periodization of oil production, (3) theory of competition and formation of rent in the oil industry, and (4) theories of the oil value and market prices.

It will be demonstrated that the oil crisis of 1973-74 is the outcome of an increase in the value magnitude of the least productive oil region of the world. This implies that the production and formation of value and the market prices have become integrated within the structure of the international oil industry. It will also be shown that an increase in the *socialization of production,* in the oil industry in conjunction with the emerging forces of competition through reciprocal relations of individual capitals, has ultimately led to the formation of differential rates of profit at the global level. The material basis of these differential profit rates relates to the existence differential productivities in the oil industry. In addition, due to the predominance of a particular set of *property relations* in the production of oil, these differential productivities are regularly being transformed into *differential oil rents* on a *permanent* basis. It will be maintained, therefore, that it is due to competition, rather than monopoly, that the least and the most productive oil producing regions of the world are objectively forced to stand in reciprocal relations that would, in turn, produce differential oil rents of various magnitude. In this connection, it will be pointed out that it is the *regulating capital* invested on the least productive oil region that tends to determine the market value of crude oil internationally.

This volume is divided into two parts. Part one includes the first three chapters, following the introduction, that would discuss the historical background and critique the theories of the oil crisis. Part two contains the six remaining chapters that would establish a theoretical framework for the analysis of the oil industry in general, and focus on the condition of the US oil production, in particular.

Chapter one presents a survey of the existing theories of the oil crisis. These theories are divided into three categories: (1) traditional theories of the oil crisis, (2) Dependency theories of the oil crisis, and (3) conspiracy theories of the oil crisis. Appropriate critique of these theories will be offered in chapter two, where we conclude that none of the existing theories are capable of explaining the *cause* of the oil crisis of 1973–74.

Chapter three examines the notion of oil property relation, and the historical development of petroleum production in the Middle East. Here, an attempt has been made to periodize the entire history of oil into three distinct intervals. Chapter four discusses the importance of raw materials in the production of value. This includes a brief section on the evaluation of the division of resources into exhaustible and renewable categories.

Chapter five deals with the theories of rent and their critique. Here, we attempt to formulate a theory of oil rent that would explain the characteristic of property relations that exist in the oil industry. The subject of chapter six is the evaluation of the notion of *pure competition* and its relevance to the actual process of competition in capitalism. Chapter seven deals with the empirical examination of the US oil production. Here, it will be established that the United States and the Middle East are the least and the most productive oil regions of the world, respectively. In addition, it will be demonstrated that there has been a significant decline in the productivity of the value-determining US oil fields since the beginning of the 1970s.

Even though in chapter seven we have shown that the increase in the price of oil during the period of 1973–74 is closely associated with the decline of productivity of investments that are made on the previously developed US oil fields, nevertheless, we use a different approach to prove the same points in chapter eight. Therefore, a portion of this chapter, regardless of its mode of presentation, i.e., econometrics, reveals nearly the same messages that are produced previously. In the remaining portion of this chapter, we attempt to measure the changes that have occurred in the size of oil royalties associated with the Middle East and the US oil regions. To conclude our study, the last chapter will put together the elements of an alternative theory of the oil crisis. Here, contrary to 'conventional wisdom', we maintain that the increase in the oil royalties of the Middle East was *not the cause* but the consequence of the oil crisis.

INTRODUCTION

The subject of the present study is the oil crisis of 1973–74, which was triggered by the oil embargo of October, 1973. This embargo was initiated by a group of Arab oil-exporting countries in response to the United States support of Israel during the Arab–Israeli War of 1973. Soon after the oil embargo, oil spot prices soared to levels which were unprecedented in the history of oil production. The range of bids for the auctioned crude oil by Iranian, Libyan, and Nigerian governments was somewhere between $17 to $24 per barrel in December 1973. Meanwhile, during the same period, 'posted prices' increased fourfold, and consequently the oil income of the countries in the Organization of Petroleum Exporting Countries (OPEC) was increased substantially.

Even though the period of the oil embargo was brief and supply conditions returned to normal at the end of 1973, the price of crude oil remained permanently at its new level.[1] Even the worldwide recession of 1974–75 did not have a significant impact on the oil price structure.

The oil crisis of 1973–74, in addition to its economic impact, had a strong political impact upon the public at large in the United States, as well as upon the people of other countries. In the United States alone we witnessed tremendous concern over the issue of the oil crisis following the Arab oil embargo. The public opinion polls of this period, however, show mixed results. For instance, a US Gallup poll taken during the month of December, 1973, in response to the question 'Who or what do you think is responsible for the energy crisis?', registered the following results:[2]

---

1 Contrary to a number of writers on this subject, we are concerned with the price of crude oil as a final product, rather than the value of a barrel of oil in terms of the price of its derivatives and by-products. The latter treatment, in our opinion, confuses the process of price formation in one with those of others. Therefore, it would be erroneous to include the consuming government charges that are usually collected through taxing of by-products as part of the price of crude oil. Contrary examples can be seen in J.M. Chevalier, 'Theoretical Elements for an Introduction to Petroleum Economics', in *Market, Corporate Behavior, and the State*, edited by A.P. Jacquemin and H.W. de Jong (The Hague: Martinus Nijhoff, 1976) and Mohssen Massarrat, 'The Energy Crisis', in *Oil and Class Struggle*, edited by Petter Nore and Terisa Turner (London: Zed Press, 1980).

2 Printed in James W. McKie, 'The United States', *DAEDALUS*; Fall, 1975, p. 85.

	Percent
The oil companies	25
The federal government	23
The Nixon Administration	19
US consumers	16
Arab nations	7
Big business	6
Leaders 'playing politics'	4
US exporting	3
There is *no* shorgage	6
Misc./no opinion	19

Despite this ambiguity, the reaction of the US government and 'emergency measures' taken by the Nixon Administration were alarming. Nixon and other members of his administration tried hard to convince the public that the shortage was real and the Arabs were responsible.

The oil crisis of 1973-74 was also a turning point which triggered, and is still stimulating, a monumental amount of literature. The theorists of the oil crisis have written on various aspects of the problem, including the cause or causes of the oil crisis. However, in examining the direction of the analysis in this literature, it becomes apparent that the theorists proved more able to examine the impact of the crisis than to determine its causes. Consequently, at the present time there is a wide gap in the existing literature which stems either from a failure to distinguish the cause from the consequences or the ignorance of the authors to recognize the genesis of the oil crisis in the first place. The present study is an attempt to bridge this existing gap by developing a *theory of the cause.*

In order to develop an alternative theory that will be able to explain the cause of the oil crisis of 1973-74, one needs, first of all, to examine all the existing major theories of the oil crisis. In this connection, the concrete events associated with the oil crisis within the context of the international oil industry have to be carefully identified and studied. Finally, the cause of the oil crisis must be distinguished from the effects and consequences of the crisis within the context of the world economy.

We hypothesize that *the cause of the oil crisis of 1973-74 is the increase in the regulating capital costs associated with the aged oil fields located in the lower 48 states of the United States.* This recognition is crucial because it presupposes the process of integration of production and the formation of value at the level of the international oil industry. This process necessarily relates to parallel developments in the structure of oil property relations that necessitate the formation of differential oil rent at the international level. As a result, there is a tendency for the social value of the entire oil industry to be formed in association with the least productive regions. Therefore, any significant changes in the cost structure of the least productive oil regions must necessarily lead to a change in the regulating capital and the value structure of oil production globally. This implies that there is a need for restructuring of capital. That is what we mean by the *'oil crisis'*.

We further hypothesize that the US oil region has been the least productive area both before and after the crisis. Thus, a further decline in the productivity of US oil must lead to a substantial increase in the magnitude of market value for the entire oil industry.

The method of analysis of the present study combines an investigation of the implications of Marx's *theory of value*, in conjunction with an historical periodization of the oil industry in the Middle East and the United States. At the same time, the conditions of global oil production will be studied in order to establish an organic relation between the conditions of oil production in the United States and the Middle East, in order to demonstrate the significance of the oil crisis of 1973–74. The selection of the United States and the Middle East oil regions in this study is for the purpose of comparison of the least and the most productive areas of the world and their corresponding productivity differences. These productivity differences, we shall argue, are the basis of the existing differential oil rents in the oil industry.

We intend to demonstrate that the oil crisis of 1973–74 resulted from a decline in the productivity of development capital expenditures in the US oil industry. This hypothesis is based upon the phenomenon of organic integration of oil production at the global level, and, therefore, the necessity of price unification since 1970. Such an integration relates various regions of the world to each other through competition. As a result, competition of the most and least productive oil regions will necessarily lead to formation of differential oil rents.

Thus, the oil crisis of 1973–74 is the result of the emergence of a newly-formed market value for the entire international oil industry, which in turn evolved through the determining impact of the *regulating capital* invested in the least productive oil region of the world, i.e., the aged oil fields of the United States. This has to do with the changing structure of capital investment in conjunction with the existing patterns of land ownership and private leasing practices in the US oil industry which has ultimately led to the conditions of higher market value for crude oil. Consequently, the effect of the above crisis is the upward change of the 'posted price' of crude oil in the Middle East and elsewhere.

The oil crisis, therefore, is the result of the emergence of new values, oil rents, and ultimately market prices associated with this newly-developed situation in the international oil industry.

PART ONE

HISTORICAL BACKGROUND AND CRITICAL EXAMINATION OF THE THEORIES OF THE OIL CRISIS.

CHAPTER I

AN ASSESSMENT OF THE OIL CRISIS: SURVEY OF PREVAILING VIEWS

The oil crisis of 1973-74 was an important economic as well as political event that has produced controversy after controversy since its initial outbreak. The arguments over the nature, cause, and consequence of such a crisis raised a great deal of debate during the last decade. Also, due to the diversity of opinions and the number of unsettled questions, there seem to be as many theories about the oil crisis as there are theorists in this field of inquiry. However, it is helpful to distinguish the common threads of prevailing views and arguments on the above subject, in order to be able to recognize the existing theoretical lines and schools of thought.

In this chapter, the prevailing views on the oil crisis of 1973-74 are divided into three main categories: (1) Traditional Theories of the oil crisis; (2) Dependency Theories of the oil crisis; and (3) Conspiracy Theories of the oil crisis.

1. The Traditional Views of the Oil Crisis

The economists under this category raise quite an extensive spectrum of arguments about the oil crisis of 1973-74. Their analyses often contain references to oligopolistic structures of the oil companies, the collective decision making of OPEC, and the operation of supply and demand at the international oil market. Many disagreements do exist among these theorists. Nevertheless, they all approach the oil crisis of 1973-74 more or less in the same fashion, from the same theoretical and methodological standpoint.

For instance, Raymond Vernon argues that the energy crisis is the result of the low cost of crude oil relative to other sources of energy. Consequently, he asserts that as the national incomes of consuming nations increased during the last two decades, their demand for energy increased correspondingly. But since crude oil was cheaper than other sources, demand for crude oil grew faster than the demand for other sources, and this in combination with the supply interruption resulted in crisis[1]

To know how inexpensively the crude oil entered into the international market and from what oil regions of the world it was extracted, one needs to study the history of oil production in the Middle East after World War II. During this time, the need for the quick reconstruction of Western Europe, on the one hand, and the existence of cheap sources of oil in the

Middle East, on the other, led to gradual displacement of the coal industry in Europe. This substitution, as R. Prodi and A. Clô indicate, resulted in (1) rigid controls over the coal market, (2) a lack of a common energy policy and separate rules for the oil and coal markets, and (3) a lack of adequate legislative rules for the limitation of entry of multinational oil companies into the European market.[2] Therefore, the European market was flooded with cheap Middle East oil and gradually the dependence of European economies on imported oil was established.

In addition, it is argued that the entrance of the so-called independent oil companies during the 1950s and 1960s was another source of the crisis.[3] This problem is the vulnerability of 'independents' in negotiation with OPEC which, as the theorists of this category argue, 'broke the deadlock' of negotiation over the posted prices and the condition of production.[4] Thus, the pressure exerted upon Occidental Petroleum by Libya during the 1970 negotiation over the extent of the increase in the posted prices, and the percentage of 'profit sharing', coupled with restrictions in oil production under the threat of nationalization, are often considered to be the causes of the oil price increase.[5]

Beyond these factors, Edith Penrose also points out that OPEC was gaining bargaining power vis-à-vis all the oil companies. In addition, the growing dependence of the United States on Middle Eastern oil in general, and the Arab oil in particular, was added to the background of conflicts over the establishment of the 'State of Israel'. Coupled with inflation in industrial countries and worldwide confusion in the international monetary system, these conditions resulted in further disarray.[6]

A similar view on the cause of the oil crisis is presented by George Lenczowski. He states that 'the oil countries' position of weakness and the corresponding strength of the companies could be attributed to three interrelated factors: the existence of a buyer's market, the self-sufficiency of the United States as a producer and consumer of oil, and the dominant position of the major oil companies. All three factors were undermined or reversed in the early 70s'.[7]

Since the absence of self-sufficiency of oil production in the 70's the United States is a common element in most of these explanations, it seems relevant to find out how the United States has become an importer. In the opinion of James McKie, the high economic growth rate of the U.S. economy during the 1960s led to the increase in oil consumption. This, coupled with the fall in the real price of oil, resulted in higher demand for crude oil consumption. The decline in the price of oil was so sharp, he says, that in 1969 it registered about 10 percent below what it had been 11 years before. Thus, the dependence of the U.S. economy on this source of energy in general, and foreign oil in particular, has increased.[8]

John Blair also places the oil crisis of 1973-74 in its historical context by referring to the first and second 'Suez crisis' and makes comparisons between them and the Arab oil embargo of 1973. He states that during the 'first Suez crisis' in 1957, the stocks of oil were already built. Thus, there was no price increase. Likewise, during the 1967 Arab–Israeli War, there was a tremendous excess supply in the United States that prevented the impact of supply interruption in Europe and elsewhere.[9] However, it turned out that prior to 1973, the excess supply of the United States gradually vanished and the U.S. economy became dependent on imported oil.[10] Blair then concludes that 'the "energy crisis" is thus the logical consequence of the limited and declining size of U.S. reserves'.[11]

With regard to the structure of the U.S. oil market, John Blair believes that due to the nature of industry, *price stability* requires almost total control over the market. As a result, he argues, it takes only a relatively small volume of 'uncontrolled' crude oil auctioned at the spot market to disrupt price stability.[12]

The final element of the traditional explanation turns away from supply and demand and focuses on expectations. Dasgupta's and Heal's explanation about the nature of the oil crisis is very much the result of the methodological underpinnings of their analysis. They write:

> Nothing in the post-1970 experience seems inconsistent with the trend established in the period just before 1970. . . This lack of any obvious alteration in the economic circumstances affecting the market at the turn of the decade forces us to conclude that the changes must have been in expectations, in the way people perceived economic realities *rather than in those realities themselves*.[13]

These authors consider a number of factors that might have contributed to changes in people's attitudes. One factor is the changed position of the United States in the international oil market from a self-sufficient country to an oil importer. This situation, they said, resulted in recognition of the OPEC countries and the extent of their bargaining power in the international market. Another factor which induced the change in attitude of the actors in the market during the early part of the 1970s is said to be the growth of 'neo-Malthusian ideology' and the resultant idea of the preservation of resources. The last factor is considered to be the change of the financial position of the oil-producing countries and their demand for further revenue, which in turn can be remedied, either by production or price increases. Clearly, the second option was chosen in the early 1970s.[14]

It should be pointed out that in fact there are considerable degrees of differences of opinion among the theorists of this category. These differences are theoretical as well as political in nature. However, the common denominator of the works of the theorists of this category is their reference to (1) the primacy of demand and supply in regulating market conditions, (2) the conventional oligopolistic theory of the firm, and (3) the traditional view of *cartel* associated with OPEC. Therefore, the categorization of Dasgupta and Heal with the theorists, such as John Blair, should not give the impression that all their theoretical viewpoints are the same.

Dasgupta and Heal have also made some observation about the level of oil prices before and after the oil crisis. They relate the prevailing oil prices of the 1960s and 1970s by the mechanism of compound rate of return. This rate of return is the return on industrial investment or the 'opportunity cost' of capital. The above authors either move from the 'low' oil prices of the 1950s to the 'high' oil prices of the 1970s by compounding the 1950s prices, or alternatively relate the prevailing levels of oil prices in the 1970s to those of the 1950s or 1960s by discounting the 1970s prices by the so-called opportunity cost of capital. Using this method, one cannot argue that the post-embargo prices are too high or the pre-embargo prices too low. Instead, one must discover a positive range between the level of marginal extraction costs and that of competitive equilibrium prices. According to Dasgupta and Heal, the price level is connected with the range of such margins. But the 'Calculation of this margin would [of course] require a great deal of information about future supplies and demands, and about the effects of technological changes on the productivity of oil.'[15] In conjunction with the above analysis, Dasgupta and Heal suggest that if the price of oil had risen by 10 percent from the year 1950, the 1974 price would have increased to somewhere in the neighborhood of $10 per barrel.[16]

The above authors thus do not venture to comment as to whether or not oil prices were too low or too high prior to and after the 1973–74 crisis. Curiously, what remains is to comment on the *rapidity* of the price change during the 1973–74 period. They observe that such a rapid change in price resulted in the monumental problem of 'adjustment' at the international level. However, they attribute the intensity of such a rapid fluctuation in price to the absence of 'forward market' in crude oil which could have induced a gradual upward change in the trend of oil prices. Thus, the conclusion by Dasgupta and Heal emphasizes the importance of 'better information' about future markets for the purpose of a smooth transition from one set of prices to another. As a result, the above arguments lead to the statement that rapid price increases in general and the rapid change in the price of oil in 1973–74 cannot occur if adequate information is available.[17] The final result of such a view is that the crisis could have been prevented if there had been a '*forward market*' in crude oil.

To summarize the arguments made by the economists and theorists of this category, i.e., the Traditional Theory of the oil crisis, one needs to put together the combination of emphases for an explanation of the oil crisis of 1973–74. First, the 'law of supply and demand', without any connection with the law of value, is emphasized. Second, the element of monopoly and 'ability' on the part of OPEC to set 'prices' at will is considered to play a decisive role. Thirdly, the dependence of the U.S. economy on foreign oil, especially the OPEC oil, which in turn created a severe momentary shortage and threatened 'the supply security', is considered to be a determining factor. Finally, it is often said that it is the suddenness of the price change and problem of adjustment that resulted in the crisis, and if there had been the possibility of a smooth transition, the oil crisis would have been avoided. From the methodological standpoint, the crisis either explicitly or implicitly is considered to have resulted from the change in the perception of the actors involved in the oil market, rather than being the product of the changed realities of the time.

2. *Dependency Theories of Oil Crisis*

The Dependency Theories of the oil crisis of 1973-74 are deeply rooted in *Dependency Theory* in general. However, the Dependency Theory has been the subject of intense debates during the last decade, and the present study neither claims nor intends to evaluate all the issues involved.[18] The task here is to reproduce the most specific arguments of the above view which relate to the oil crisis of 1973-74. For the sake of clarity we will concentrate on Norman Girvan's evaluation of the oil crisis of 1973-74.[19] Nevertheless, there are other authors, such as Michael Tanzer and Joe Stork, who have made significant contributions to this subject from the standpoint of Dependency Theory.[20]

Norman Girvan points out that the oil crisis of 1973-74 was the manifestation of the crisis in the world capitalist economy which came about with 'the OPEC offensive of 1970-73'. His main argument in this explanation of the oil crisis of 1973-74 revolves around the concept of the 'domination-dependence relationships that characterize the contemporary world system and tensions, reactions, and conflicts generated by them'.[21]

Girvan thus forms a power-relations framework to explain the sudden fourfold increase in the 'posted price' of crude oil in 1973-74. Such a power struggle in the context of world economy is said to be the manifestation of *economic nationalism* of 'Third World' countries in order to gain a 'nearly equal role' in their relationship with the industrial powers of the center.[22] The international capitalist system in the above sense is the system of 'development-underdevelopment-domination-dependence, and center-periphery'.[23]

As a result, such a crisis is seen to be the consequence of struggle for 'national economic sovereignty' *on the part of the primary-producing countries* in general and OPEC in particular.[24] That is why for Girvan, and for the entire group of Dependency theorists, the 'OPEC offensive' is the cause of the oil crisis. This situation, in fact, is supposed to be an economic revolt on the part of oil-exporting countries, a revolt against *unequal exchange* between center and periphery, a revolt against the exchange of cheap raw materials exported from the 'Third World' countries for imported industrial commodities and finished products with high prices.[25] Therefore, the phenomena of 'equity participation' and joint ventures are seen as a trend toward sovereignty on the part of the OPEC and a tendency toward more equal exchange at the international level.

With regard to the price of crude oil after the embargo, opinion is divided within the Dependency school. One group of Dependency theorists argue that the pre-embargo prices were extremely low. After the embargo and the price increase, (see Table 1), it is maintained that unequal exchange has been limited and trade between the oil-exporting and industrial capitalist countries is now carried out under more favorable conditions for OPEC. Nevertheless, there is disagreement on the extent of the 'equality of exchange' in the oil trade among the theorists of this category.[26]

Other dependency theorists argue that because of the high rate of inflation in the industrial countries of the center, the terms of trade for OPEC soon deteriorated. As a result, the effect of the increase in the level of oil prices was gradually neutralized.

A more detailed perspective, which uses the concept of Third World nationalism is provided by Michael Tanzer. He believes that three interrelated events together led to the emergence of the 'energy crisis'. These events were (1) increased competition among oil companies, causing a decline in the price of oil, which in turn resulted in the decline of profit rates of the international oil industry; (2) environmental concerns and the uncertainty over the type of energy resource which might be dominant in the future; and (3) 'increasing nationalism' on the part of the OPEC, which in turn led the major companies to become less dependent on them as a source of oil.[27] As a result, Michael Tanzer argues, 'once the big oil companies acquired vast quantities of non-oil energy resources, they had a major incentive to try to drive up the price of all energy resources. One of the most important tactics in achieving this has been the creation of an "energy crisis" atmosphere throughout the world, and particularly in the United States'.[28]

It seems that the above version of the *oil crisis* is combining the notion of so-called economic nationalism of OPEC with the concerted efforts of major oil companies to raise the oil price, once their *monopoly position* over international oil production is restored. The characteristic of this monopoly situation, however, revolves around monopolization of the alternative sources of energy in order to counteract the competition created by smaller companies which entered the market during the period of pre-embargo. Thus, the 'energy crisis' is, first of all, the result of the emergence of monopoly conditions in the international oil industry.

Table 1

Changes in Government 'Take': Light Arabian Crude
(dollars per barrel)

	Nov. 1965	*July 1973*	*Jan. 1974*
1. Operating Cost	$ .10	$ .10	$ .10
2. Royalty[a]	.22	.38	1.45
3. Subtotal	.32	.48	1.55
4. Posted Price	1.80	3.01	11.65
5. Less Cost and Royalty (3)	- .32	- .48	- 1.55
6. Subtotal	1.48	2.53	10.10
7. Tax Paid	.74[b]	1.40[c]	5.55[c]
8. Govt. 'Take' (7+2)	.96	1.88	7.00

a at 12.5%
b at 50%
c at 55%

Source: John M. Blair, *The Control of Oil* (New York: Pantheon Books, 1976), p. 269.

Second, it coincided with the economic nationalism of OPEC. Consequently, the oil crisis is not caused by an actual shortage of crude oil, but instead initiated for the sole purpose of increasing its price.

Thus, it seems that the oil prices, both before and after the embargo, were set by the extent of monopoly in the international oil market. According to the above interpretation, during the 1960s, with the entrance of additional companies into the market, the monopoly position of the *majors* eroded and the price of oil declined. However, after the monopoly of major oil companies over other sources of energy was secured, competition suddenly disappeared and the era of monopoly pricing started. Therefore, the basis of the above interpretation, aside from the 'economic nationalism' of the producing countries, rests upon the so-called quantity theory of competition.[29]

It should be noted that some theorists would argue that the 'energy crisis' was the oil companies' tactic. Michael Tanzer is perhaps one of them.[30] Others, such as Norman Girvan, place an emphasis on the notion of 'economic nationalism'.[31]

Nevertheless, the Dependency view of the oil crisis is based upon the general theoretical framework of Dependency theory. The notion of *unequal exchange* is either explicitly or implicitly incorporated into the analysis of the oil crisis of 1973-74 by most writers in this category. 'Economic nationalism' is said to be the motivating force behind the sudden increase in the oil prices, prior to and during the embargo period. The process of price determination is conditioned by the existence of monopoly in the industry. Therefore, the law of supply and demand, subject to such a monopoly situation, determines prices. Another contributing factor to price determination during the early 1970s is the OPEC demand for higher prices. Called by some authors 'the OPEC offensive', this refers to the 'posted price' hike of 1973-74.

3. The Conspiracy View of Oil Crisis

This view is based on the idea that the U.S. government, in collaboration with the international oil companies and OPEC, intentionally brought about the oil crisis of 1973-74. As a result, the focus of the above argument is on the differential impact of the fourfold increase of the oil prices on the U.S. economy, on the one hand, and on the economics of Western Europe and Japan, on the other hand. More explicitly, it is said that even though the U.S. balance of payments was adversely affected by the sudden increase in the price of oil, the damage was negligible compared with what happened to the balance of payments of the Western European and Japanese economies. Table 2 shows the balance of current accounts of the United States, Japan, and a number of Western European countries, during the period of 1970-75.

Table 2

Current Accounts of Selected Countries
(in billions of dollars)

	1970	*1971*	*1972*	*1973*	*1974*	*1975*
United Kingdom	1.5	2.5	0.2	-3.1	-9.0	-6.5
France	- .2	.5	.3	- .7	-7.5	-6.3
Germany	1.0	.9	1.0	4.5	9.0	6.0
Italy	.8	1.9	2.0	-2.4	-8.3	-5.8
Japan	2.0	5.8	6.6	- .1	-4.8	0
Netherlands	- .5	- .2	1.0	1.8	1.5	1.5
Belgium	.7	.8	1.4	1.4	.3	0
Denmark	- .5	- .5	- .1	- .5	-1.2	-1.2
United States	.4	-2.8	-8.4	.5	-3.3	-7.5

Source: Committee on Banking and Currency, House of Representatives, 93rd Congress, 2nd Session, *Balance of Payments Adjustment to Higher Oil Prices: Managing the Petrodollars Problem,* (December 1974), U.S. Government Printing Office, Washington: 1975, p. 27.

The changes in the balance of current accounts of a selected number of Western European countries, Japan and the United States both for pre-embargo and post-embargo periods, are illustrated in Table 3. It should be noted that West Germany is excluded from the analysis because of the fact that its balance of current accounts improved during the crisis. The countries selected are the United Kingdom, France, Italy, Denmark, Japan, and the United States. During the period of 1971-72, as Table 3 shows, the total current accounts of the first five countries registered a $1.2

billion decline. For the same period, the United States balance of current accounts showed a $6.6 billion decline. During the period of 1972–73, the current accounts of the above selected countries of Western Europe and Japan together showed a $15.8 billion decline as opposed to an $8.9 billion gain by that of the United States. Meanwhile, the change in the balance of current accounts of the above European and Japanese economies during the period of 1973-74 showed a further decline to the level of –$24.0 billion, while the United States declined only to –$3.8 billion. These relative declines in the balance of current accounts of the countries selected for illustration is the basis for the conspiracy theorists' argument that the U.S. government must have initiated the process of the price hike in order to hurt the above-mentioned countries.

In addition, it is often maintained that the major oil companies, with OPEC, deliberately increased prices and coordinated their action with the U.S. government. Thus, the oil crisis of 1973–74 is seen to reflect the domination of the United States over the globe, a domination which since World War II was obtained by under-cutting such countries as the United Kingdom.[32]

Table 3

Changes in the Balance of Payment of a Selected Number of Industrial Countries Before and After the Arab Oil Embargo
(in billions of dollars)*

	1971–72	*1972–73*	*1973–74*	*1975–75*
United Kingdom	– 2.3	– 3.3	– 5.9	+ 2.5
France	– 0.2	– 1.0	– 6.8	+ 1.2
Italy	+ 0.1	– 4.4	– 5.9	+ 2.5
Denmark	+ 0.4	– 0.4	– 0.7	0
SUBTOTAL	– 2.0	– 9.1	–19.3	6.2
Japan	+ 0.8	– 6.7	– 4.7	+ 4.8
TOTAL (Japan and Western Europe)	– 1.2	– 17.8	–24.0	+11.0
United States	– 6.6	+ 8.9	– 3.8	– 4.2

*Calculated by the author from Table 2.

This conspiracy theory of the oil crisis is mostly popular in West European economic and political circles. For instance, even prior to October 1973, before the initiation of the oil embargo, the London *Economist* of July 1973, while commenting on the OPEC negotiations of 1970–71, stated:

> . . . the suspicion remains that the American State Department caved in too easily, though to ascribe this to simple blundering is not convincing either. . . One suspicion is that some concern from a higher level of politics affected the way the Americans acted. . . According to it, the Americans gave in to OPEC so readily because they saw increased oil prices as a quick and easy way of slowing down the Japanese economy, whose exports were bothering America mightily at the time and which would be more hurt by rises in oil prices than any other nation.[33]

In the United States, this view produced a great deal of controversy in a variety of economic and energy conferences, forums, and conventions. A complete account of the Conspiracy Theory of the oil crisis was formally publicized in an article in *The Washington Post* of July 10, 1977.[34] This article declared that:

> It sounds off to American ears, but sceptical Europeans have been talking for years about the economic advantages and the increased political leverage that America may have gained from OPEC. In some quarters, there is even a suspicion that U.S. diplomats implicitly encouraged the OPEC nations to jack up their oil prices *in order to enhance the American position in the world.*[35]

The authors were perhaps conveying the content of a report given by Jack Anderson and Les Whitten, which was published a few days earlier in the same newspaper. Anderson and Whitten claimed that they 'Now have evidence that Richard Nixon and Henry Kissinger could have prevented the spectacular oil price increase that has dislocated the world economy and sent all prices spiraling.[36]

A characteristic of the Conspiracy view of the oil crisis is centered around the rivalry of the United States with the European and Japanese economies in the context of trade and balance of payments. The sudden increase in the price of crude oil in 1973–74 was the result of the coordinated efforts of the U.S. government with the major oil companies and the most accommodating members of the OPEC to increase the price of crude oil. Although the United States has become a net importer, the effect of the price hike was certainly more pronounced in Japan and Europe, where almost all the oil consumed is imported oil. Another feature of the Conspiracy Theory of the oil crisis is the political nature of its theory of price determination; the price hike is not due to the objective determinants of supply and demand, but rather is the result of a coordinated exercise of *political will.* The plausibility of this thesis rests on the assertion that a great deal of harmony exists between the oil companies and OPEC and that the primary conflict in the world economy is between the United States, on the one hand, and Europe and Japan, on the other.

NOTES

1. Raymond Vernon, 'An Interpretation', *DAEDALUS:* Fall, 1975, pp. 1-14.
2. Romano Prodi and Alberto Clô, 'Europe', *DAEDALUS:* Fall, 1975, p. 92.
3. Raymond Vernon, 'An Interpretation', *DAEDALUS,* 'Fall, 1975, p. 4.
4. *Ibid.*
5. Edith Penrose, 'The Development of Crisis', *DAEDALUS,* Fall, 1975, p. 41.
6. *Ibid.*, p. 39.
7. George Lenczowski, 'The Oil Producing Countries', *DAEDALUS,* Fall, 1975, p. 60.
8. James W. McKie, 'The United States', *DAEDALUS,* Fall, 1975, p. 73.
9. John Blair, *The Control of Oil* (New York' Pantheon Books, 1976), p. 3.
10. *Ibid.*, p. 3.
11. *Ibid.*, p. 4.
12. *Ibid.*, pp. 52-53.
13. P.S. Dasgupta and G.M. Heal, *Economic Theory and Exhaustible Resources* (Cambridge: Cambridge University Press, 1979), pp. 444-45 (emphasis added).
14. *Ibid.*, pp. 445-46.
15. *Ibid.*, pp. 451-52.
16. The authors propose that 10 per cent per annum is the rate of return that was obtainable by industrial investment during the same period (*Ibid.*, p. 451).
17. *Ibid.*, p. 454.
18. For the study of Dependency Theory, among others, one can examine various issues of *Latin American Perspectives,* especially those in vol. I, no. 1 (Spring, 1974); vol. VI, no. 2 (Spring, 1979); and vol. VIII, no. 3 and no. 4 (Summer and Fall, 1981). Also, for a lucid critique of unequal exchange, see Charles Bettleheim's 'Theoretical Comments In A. Immanuel, *Unequal Exchange* (New York: Monthly Review Press, 1972).
19. Norman Girvan, 'Economic Nationalism', *DAEDALUS,* Fall, 1975, pp. 145-58.
20. Michael Tanzer, *The Energy Crisis: World Struggle for Power and Wealth* (New York: Monthly Review Press, 1974); Joe Stork, *Middle East Oil and the Energy Crisis* (New York: Monthly Review Press, 1975).
21. Norman Girvan, 'Economic Nationalism', *DAEDALUS,* Fall, 1975, p. 145.
22. *Ibid.*, p. 145.
23. *Ibid.*, p. 147.
24. *Ibid.*, p. 146.
25. *Ibid.*, p. 148.
26. See for instance, Norman Girvan and Joe Stork for comparison.
27. Michael Tanzer, *The Energy Crisis: World Struggle for Power and Wealth* (New York: Monthly Review Press, 1974), pp. 31-32.
28. *Ibid.*, p. 34.
29. See Anwar Shaikh, 'Political Economy and Capitalism: Notes on Dobb's Theory of Crisis', *MIMEO* (New York: New School for Social Research), February 1978; John Weeks, 'Marx's Theory of Competition and the Theory of Imperialism', *MIMEO,* November 1977, Department of Economics, The American University.
30. Michael Tanzer, *The Energy Crisis: World Struggle for Power and Wealth* (New York: Monthly Review Press, 1974), p. 34 and p. 122.
31. Norman Girvan, 'Economic Nationalism', *DAEDALUS,* Fall, 1975, pp. 145-58.
32. Michael Tanzer, *The Energy Crisis* (New York: Monthly Review Press, 1974), pp. 17-18; Michael Tanzer, *The Race for Resources* (New York: Monthly Review Press, 1980), pp. 108-10; and Joe Stork, *Middle East Oil and the Energy Crisis* (New York: Monthly Review Press, 1975), Chapter 2.

33. Quoted in Michael Tanzer, *The Energy Crisis: World Struggle for Power and Wealth* (New York: Monthly Review Press, 1974), p. 135.
34. William Greider and J.P. Smith, 'A Proposition: High Oil Prices Benefit U.S.', *The Washington Post,* (July 10, 1977), p. 1.
35. *Ibid.*, p. 1 (emphasis added).
36. Jack Anderson and Les Whitten, 'Kissinger Allowed Oil Prices to Soar', *The Washington Post,* (July 7, 1977), p. D15.

CHAPTER II

A CRITIQUE OF THE PREVAILING THEORIES OF THE OIL CRISIS

During the course of the previous chapter, various theories of the oil crisis associated with the three broad categories of theories were reviewed without any attempt to make them the subject of critical examination. In the present chapter, however, we need to evaluate the significance of these theories which emerged during the last decade and determine the extent of their relevance to the actual process of the oil crisis.

In the first section, a critique of the Traditional Theory of the oil crisis will be presented. A critique of the Dependency Theory and the Conspiracy Theory of the oil crisis will be discussed in sections 2 and 3.

1. *Critique of Traditional Theories of the Oil Crisis*

One of the common errors often committed by the theorists in this category is the lack of distinction between the causes and the effects of the oil crisis. As we have discussed in the previous chapter, the three basic elements of the oil crisis as expressed by a majority of the theorists in this category were (1) the supply conditions due to the oil embargo, (2) the ability of OPEC to set prices, and (3) dependence of the United States on the imported oil.

From the viewpoint of supply conditions, to put it in Blair's terms, the short falls were short lived after the oil embargo.[1] In other words, a temporary shortage developed that according to *conventional economic theory* should not have had any impact on the long-term equilibrium prices. However, as the history of this period vividly indicates, exactly the opposite resulted. In this manner, according to the above view the factors which established a higher floor for the post-embargo price of oil are the combination of price determination by OPEC and *dependence* of the U.S. economy on imported oil.

It should be noted from the outset that the first step in this analysis is to identify the cause or causes of the oil crisis. Despite this necessity, most theorists in this category counted a number of factors associated with the crisis, without actually developing any systematic relationship among them in order to identify the underlying causes. In the final analysis, the 'OPEC-determined' oil prices (along with the notion of the U.S. 'dependence' on imported oil) is considered to be the cause of the oil crisis by the majority of the theorists of this category.

There are, however, a number of misconceptions in this conclusion.

First of all, the mechanism of posted price and market price of oil are confused with each other.[2] Second, the nature of price determination is dependent upon the *will* of OPEC, rather than that of the objective production conditions. Third, the above conclusion implies that the United States was unable to challenge OPEC because the United States is a net importer and could not supply the world market as much as it had during the 1958 and 1967 wars in the Middle East. The question to be asked is why, considering the condition of excess supplies which prevailed following the removal of the embargo, the oil prices did not decline significantly after 1974. If the 'law of supply and demand' does not hold in this particular case, then what kind of mechanism regulates the process of price formation in this industry?

Another common feature of the theorists of the present category is that they have not developed a unique mechanism to connect the process of determining pre-embargo oil prices with post-embargo oil prices. The reason for this seems to be the lack of theoretical and historical perspective. These theorists either rely on supply and demand conditions, or expand upon the notions of monopoly, cartel and oligopoly to describe the sudden price change of 1973–74. (See Figure 1.)

We have seen that after the oil embargo ended, and even before the oil embargo, the international market was flooded with crude oil. But surprisingly, the decline in the quantity demanded due to the worldwide recession of 1974–75 did not seem to cause any substantial drop in the price of oil. In fact, after the oil embargo of 1973, a new floor for the oil prices was once and for all established. This implies that the supply and demand conditions are themselves the consequence of more fundamental changes that emerged in the international oil industry. In the final analysis the traditional theorists argue that OPEC acted like a monopoly in setting the oil prices unilaterally at its fourfold increase level. When asked why OPEC did not behave in the same manner in the previous years, most of the theorists of this category reply that in addition to the existence of monopoly, factors such as the U.S. 'dependence' on the foreign oil and the rising tides of resource conservation contributed to the crisis. The most explicit argument of this sort is expressed by Dasgupta and Heal. They simply state that all these developments were the result of changes in perception of the agents involved and not the result of the changes in the actual situation.[3]

There is a fundamental methodological problem with the above theoretical formulation: the above argument implausibly suggests that *objective realities* do not exist outside of one's subjective perception. That is why Dasgupta and Heal remark that in the absence of future markets it is difficult to know how the perceptions of buyers and sellers are formulated. However, what the 'future market' does is to project the *range* of fluctuations and not the fundamental changes that

Figure 1

*Changes in Posted Price of Saudi Arabian Crude Oil**

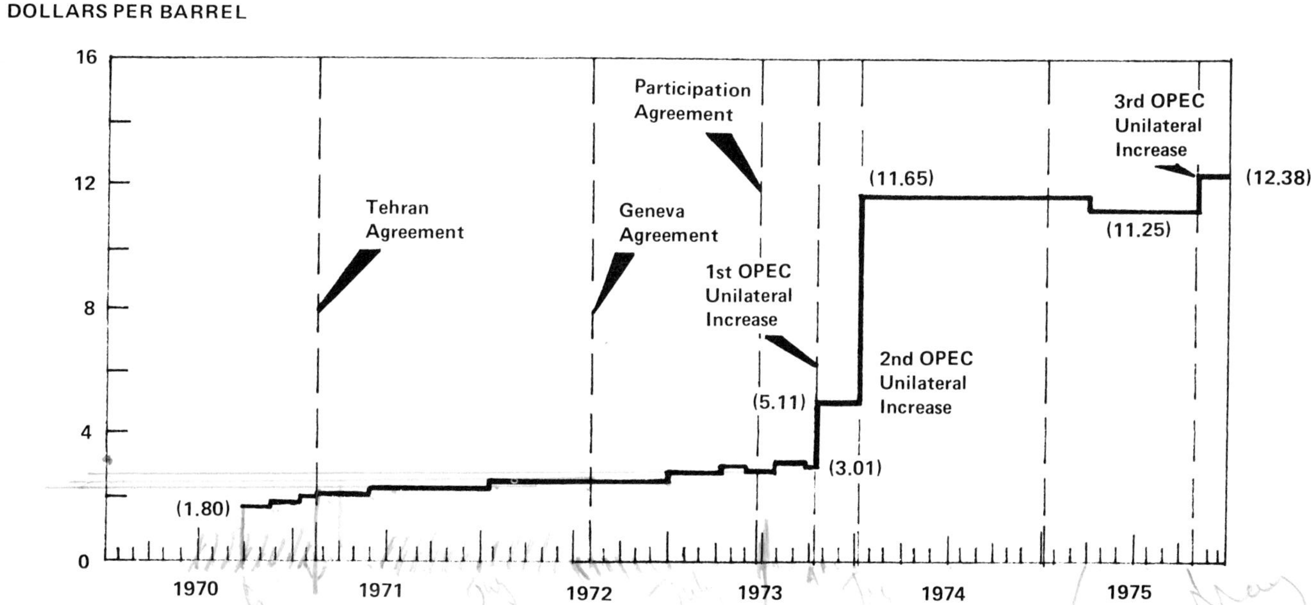

Source: John M. Blair, *The Control of Oil* (New York: Pantheon Books, 1976), p. 263.

* Arabian Light 34°

might occur in the *structure* of value formation and, ultimately, in the price formation in any one industry.

Despite the lack of recognition of the fundamental and structural changes in value formation and eventual price formation together with price fluctuations due to supply and demand, it should be acknowledged that the majority of the theorists of this category made contributions in connection with the description of events which took place during the crisis period. For instance, the description of the 'OPEC monopoly' or the notion of 'dependence' of the United States on the imported oil are true observations of the phenomenal forms of the crisis. However, what is rarely recognized is the distinction between the observable effects and the causes which would lead to them.

As a result, the above theorists were unable to theorize the process of the oil crisis of 1973–74, but at best described the consequences of the crisis and its conditions. Their descriptions of supply-demand, the 'posted price' determination of OPEC, and the dependence of the United States on foreign oil, are certainly true. What is also true is that these theorists failed to penetrate beyond the appearance of these facts in order to build a theory of the oil crisis. Instead, they more or less described the situation without understanding its underlying cause.

2. *Critique of Dependency Theories of the Oil Crisis*

An appropriate critique of Dependency theories of the oil crisis obviously has all the essential ingredients of a critique of the Dependency School in general. However, the purpose of the present section is not to evaluate Dependency Theory, but to analyze briefly the specific features of the Dependency theories of the oil crisis. The major argument expressed by most of these theorists depends on a hypothesized 'OPEC offensive' against the imperialist countries in order to achieve self-determination and sovereignty over their national resources.[4] This 'offensive', however, is said to be a reaction to the prolonged relations of domination existing between the industrialized countries of the West, on the one hand, and 'Third World' countries, on the other. This domination in turn led to unequal relations and, consequently, *unequal exchange* in trade between the 'center' and 'periphery'. Some authors of this category stated that the atmosphere of the post-Vietnam era created a general political condition that allowed the dominated countries of the 'Third World', notably the OPEC nations, to launch an 'offensive'.

The most fundamental error of the theorists in this category is *eclecticism,* i.e., putting all the social, political, and economic ramifications on equal footing without an ability to identify their structural relationship in order to single out the underlying specific cause(s) of the oil crisis. For instance, the process of price formation in the oil industry is identified with the determination of 'posted prices' by OPEC. There is no consideration of value and price formation in the international oil industry, and instead, they substitute the notion that *monopoly* determines everything. It is not surprising that the Dependency view considers the oil crisis either an offensive against *unequal exchange*, or simply a *phony crisis*, since the outcome did not, in the final analysis, change the magnitude of the existing 'unequal exchange'.[5]

These theorists failed to recognize that prices are the phenomenal form of values, and value formation emerges through competition. Therefore, they made a double mistake of emphasizing the idea of *absolute monopoly* and price determination without considering a theory of value formation in the oil industry. When these theorists discuss competition, the number of firms in the industry is assumed to be large; while monopoly suggests the existence of few firms in an industry. It should be pointed out that such a presumption is a clear misconception of both monopoly and competition, since from competition to monopoly there is only a quantitative change. These theorists fail to realize that monopoly and competition in capitalism are part and parcel of a synthesis; they cannot be separated by an arbitrary consideration of the number of firms in an industry.

The oil crisis of 1973–74 is thus seen to result from the *direct political action* of OPEC. For instance, Tanzer has argued that, 'As a result of the Arab oil embargo in late 1973, the OPEC countries effectively took over the ownership of their crude oil reserves and oil pricing, while the companies became primarily suppliers of technology and markets. . .'[6] These changes seem to be of an arbitrary nature, if one accepts the arguments advanced by the theorists of the Dependency category. They are arbitrary because they are not the result of an identified mechanism of value formation and eventual price determination. Consequently, most of the theorists in this category do not propose any systematic analysis, except the proposition of the reaction to *unequal exchange*, in order to explain the oil crisis.

In this category, like others, the oil Dependency theorists do not completely agree with each other. For instance, in explanation of the oil crisis, Tanzer's emphasis is on the monopoly aspect of the international oil industry, while Girvan's primary concern is the 'OPEC offensive'. However, from the standpoint of methodology, they both represent different versions of the Dependency Theory.

Finally, some of the Dependency theorists allege that the social and political conditions of the post-Vietnam era, i.e., the consequences of defeat of U.S. imperialism, are the cause of the oil embargo and the subsequent 'OPEC offensive', during the oil crisis. Still others put together numerous factual statements, such as the U.S. political decline, more participation on the part of the OPEC countries, increased income of the OPEC countries, etc., in order to demonstrate that the Dependency theory is compelling. These arguments are controversial because their common thread is the assumption that the 'OPEC offensive' was the result of a contradiction between the masses of the 'Third World' and U.S. imperialism. In the following chapters we will prove that the above statement is a misconception and that the contradiction associated with the oil crisis has primarily emerged within the camp of capitalism at the international level.

3. *Critique of Conspiracy Theories of the Oil Crisis*

The Conspiracy Theory of the oil crisis is a conception which overemphasizes the conflict between the economies of the United States, on the one hand, and the economies of Western Europe and Japan, on the other. The center of the analysis of the theorists in this category is the impact of the sudden change in the price of oil on the balance of trade of the United States, Western European, and Japanese economies.

These theorists argue that, while the United States has lost a great deal of *money* to imported oil, Western Europe and Japan, which were obviously more dependent on foreign oil, paid more for their imported oil. As a result, the differential impact of the sudden increase in the oil prices, in fact, significantly harmed the European and Japanese positions in international trade.

The basic error committed by the theorists in the present category is the reliance on the balance of trade for such an analysis. Of course, the increase in the price of oil resulted in a more severe impact on the balance of trade of those economies which were more dependent on foreign oil. However, this outcome does not have anything to do with cause(s) of change in the oil prices, unless one believes that the price of oil is monopolistically determined i.e., by the U.S. government, oil companies, and/or OPEC.

Once again, the same problem existing in the previous theoretical formulations comes to the fore. But the difficulty here has a different dimension. The static determination of prices by the monopoly is one thing; its conspiratorial determination by a particular government or organization is another. The misconception of such a theory is that there is no economic formulation for the explanation of the process of price formation in general, and the formation of oil prices during the crisis of 1973-74 in particular.

There is no economic and political connection between the actual causes of the oil crisis and its consequences. There is only an alleged motivation for conspiracy on the part of the U.S. government in order to have a better balance of trade. Meanwhile, it is evident that within a short time after the crisis, the balance of trade of the United States declined substantially and the favorable differential impact of the oil prices on that tended to deteriorate sooner than was expected.[7]

Another major obstacle in connection with this view is the *impossibility* of its empirical verification. In other words, it is impossible to prove or disprove that in fact there was a conspiracy to create this crisis. Even if one were able to document some evidence that in fact the U.S. officials welcomed the above changes in the oil market, one cannot build a theory on such a motivation alone.

Economic crisis is not a phenomenon that can be created by one authority or another. That is why the Conspiracy Theory of the oil crisis remains a subjective theory at the lowest level. At best this theory is about the state of mind of the U.S. officials, international oil executives, and the OPEC representatives. It does not take into account the economic and political process of the oil crisis, regardless of its political and economic consequences. The core of this theory is the particular effects of the crisis. Once these effects are demonstrated, these theorists resort to speculation in order to determine the causes of the oil price change. In this sense, the above theory can be called a *speculative* theory of the oil crisis. It should be understood that we do not wish to deny the possibility of conspiracy. The existence of such a conspiracy depends, however, upon the identification of the actual cause of the oil crisis and not the other way around. To deny the validity of the Conspiracy Theory of the oil crisis, therefore, does not lead to denial of the rivalry among the industrial countries, for rivalry is a real phenomenon and an objective process. However, the lack of validity of the above analysis is due to mistaking a phenomenal form for its essence. Thus, the cause of the oil crisis is attributed to the subjective minds of individual agents by way of conspiracy.

4. *Conclusion*

We have seen that all the above theories of the oil crisis are generally subjective in nature and also speculative to various degrees. These theories have dealt with the effects of the oil crisis, and, either partially or entirely, tended to regard these effects as causes. Some of the theorists associated with Traditional theories of the oil crisis in fact deny that the crisis was an objective process. Instead, they emphasize the role of *changed perceptions.* Others in this category, while acknowledging some objective changes, do not penetrate below the surface of appearance and describe the phenomenal form of the crisis. The Dependency theory of the oil crisis has emphasized the notion of 'OPEC offensive' within the 'center-periphery' economic relations. Thus, most of the theorists of this category stress unequal exchange and the 'challenge' of OPEC. Further, conspiracy theorists have reasoned through the differential impact of the oil crisis on the balance of the trade of the United States, Western European and Japanese economies, that there was a conspiracy in the *'arrangement'* of the oil crisis among the U.S. government, oil companies, and OPEC. All of the theorists we have examined have ultimately failed to recognize that their arguments view the crisis as a *voluntary* act rather than an objective social and economic process. In addition, they have committed the error of taking the reflection of the crisis for its actual substance.

NOTES

1. John Blair, *The Control of Oil* (New York: Pantheon Books, 1976), pp. 266–68.
2. The posted price of crude oil is a mechanism for determination of the oil royalties and rent due to the OPEC countries, while the market price is the price of a single quantity of crude oil at the open markets such as the Rotterdam oil market.
3. Dasgupta and Heal, *Economic Theory and Exhaustible Resources* (Cambridge: Cambridge University Press, 1979).
4. See also J.M. Chevalier, 'Theoretical Elements for an Introduction to Petroleum Economics', in *Market, Corporate Behaviour, and the State,* edited by A.P. Jacquemin and H.W. de Jong (The Hague: Martinus Nijhoff, 1976), p. 294 and 298.
5. Andre Gunder Frank, *Crisis: In the World Economy* (New York: Holmes & Meier Publishers, Inc., 1980).
6. Michael Tanzer, *The Race for Resources: Continuing Struggle Over Minerals and Fuels* (New York: Monthly Review Press, 1980), p. 110.
7. See Tables 2 and 3.

CHAPTER III

AN HISTORICAL ANALYSIS OF THE MIDDLE EAST PETROLEUM PRODUCTION: THE EMERGENCE OF THE CONCESSION SYSTEM AND BEYOND

1. An Overview

Long before the discovery of oil, the influence and eventual domination of imperialism in the Middle East was carried out successfully, either by military force or by political means. The decay of the prevailing pre-capitalist social order in this region, along with the lack of capability on the part of the existing state structure or the ruling authorities, led the states to collaborate with the colonial powers in order to keep the existing conditions intact.

This had been more or less the prevailing political and economic condition before the *export of capital* became the mechanism of penetration into the economies of the Middle Eastern societies, and their eventual integration into the world economy. As a result, the discovery of oil in the Middle East should be seen as a stepping stone for the extension of the material basis of the social relation of *capital* into the Middle Eastern economies, as well as a key element in the eventual consolidation of neo-colonialism. For such a discovery not only led to the emergence of the modern oil industry in the Middle East, but also, as we will see in this chapter, it simultaneously led to the transformation of social, political and economic conditions in these dominated societies.

The first major oil discovery in the Middle East was made in southern Persia in 1908, following the conclusion of a concession signed between the Persian government and a British investor by the name of William Knox D'Arcy in 1901. This was followed by a tremendous number of discoveries in Iraq (1927), Bahrain (1932), Saudi Arabia (1938), Kuwait (1938), and finally Qatar (1939).[1] These discoveries, as *potential* sources of material wealth for the production of *surplus value* under *monopoly capital,* were the basis for the emergence of modern industry, and therefore, the transformation of the existing social and productive relations in the Middle East. However, prior to these discoveries, the international oil companies needed to secure the *necessary* property rights that were *essential* for the exploration, development and production of oil in that region. The obtaining of such rights has since been the subject of the existing oil concessions signed by the parties involved in the oil history.

An oil concession is a contractual agreement concluded between a private entity, firm or individual, and a particular government. This contract regulates the transfer of oil property rights and the way in which oil can be explored, developed, and extracted by the investing party or parties. Also, any substantive change in the nature of these oil concessions does necessarily lead to the transformation of the prevailing social and economic conditions of oil production in the societies involved. Hence, in order to account for such changes it is useful to distinguish three different periods in the history of the Middle East oil industry. First was the era of early oil concessions, during which the existing international oil companies made an early attempt to penetrate and expand into the pre-capitalist economies of Middle Eastern societies. This was done by way of obtaining oil concessions, i.e., gaining exclusive property rights for exploration, development, and production of crude oil in these areas. Second was the period of transition from the early to the present time during which the Middle East oil industry has been eventually transformed into the modern industry of today. Finally, there is the period in which the oil industry in the Middle East has become a full-fledged capitalist entity, integrated in the world market. In this latter period, due to the *internationalization of all circuits of capital* in the oil industry, market values, prices and oil rents are being determined within the international arena, as opposed to the regional markets. Furthermore, it is only during this period that the concept of *economic crisis* in the oil industry can be theoretically justified.

These periods can only be separated by the identification of specific economic developments pertinent to them. These economic developments in turn have to be qualitatively differentiated as the organization of oil production and the forms of property relations change. The above periods are historically identified as: (1) the early period of 1901–1950, (2) the transitional period of 1950–1970, and (3) the period in which production and formation of values, market prices and oil rents developed at the international level (since 1970).

2. *The Early Oil Concessions*

The laws of the oil concessions governing the *dominated* oil regions of the world, including the Middle East are substantially different from the leasing contracts that prevail in the United States. It should be noted that the essential characteristic of the U.S. leasing practices stems from the structure of ownership of the subsoil, which is included as a part of the ownership of land. Due to the observance of the *rule of capture*, in the United States the materials obtained from the subsoil belong to the owner of the land.[2]

The historical development of such a system, however, is the consequence of the existence of the institution of private land ownership in the United States in general, and that of oil ownership, in particular. The notion of private land ownership itself corresponds with the nature of prevailing social relations which in turn correspond to the extent of development of capitalism in the United States.

In the absence of predominant private land ownership and modern landed property in the pre-capitalist economies of the Middle East, it goes without saying that the phenomenon of the private ownership of subsoil never existed. The oil concessions were usually a contractual agreement between an international oil company or a consortium of international oil companies and oil exporting states. These concessions usually specify the *surrender* of the right of exploration, development, and production of oil, natural gas and other related substances to the oil companies. These contracts define 'the grant, the area, the duration of the concession, the payments to be made by the concessionaire and, mostly in general terms, the mutual rights and obligations of the parties'.[3]

The term *concession,* rather than oil lease, is used to indicate that at least one party involved in the contract is a government. As a result, from the legal standpoint, an oil concession is actually a mixture of both a private and public contract. Most of the oil concessions concluded during the period of 1901–1950 in the Middle East, as summarized by Henry Cattan, had the following features:

(1) They all covered the entire or a substantial part of a country and for significantly long period of time.
(2) The number of oil companies involved was fairly small.
(3) Concession terms were often uniform and simple.
(4) Royalty payment was the principal financial transaction.
(5) The financial terms were extremely moderate.
(6) There was little change in the terms and conditions of the concessionary agreements during this period. This by itself indicates that the conflict between the parties was minimal.[4]

From the perspective of the expansion of capital and spread of capitalism at the global level, the early oil concessions are an example of the initial attempt made by the oil companies to internationalize their interests. The fact that the size of acreage (land) under a single concession was so large and often included the entire country demonstrates that the primary objective did not lie exclusively in exploration of oil for immediate extraction, but in the main the acquisition of oil property rights in the entire region. Oil companies' actions do one must say, correspond with the Leninist thesis that conglomerates divided the world among themselves during the early 20th century.[5]

For instance, the concession obtained by William Knox D'Arcy in 1901 included the total area under the sovereignty of the then Persian government, with the exclusion of five northern provinces bordering Russia. This area, which was 480,000 square miles, was given to D'Arcy and later on to the Anglo-Persian Oil Company for oil exploration. The granting of such a vast amount of land in a concession was the rule rather than the exception in the Middle East. This condition, aside from the obvious economic consequences, brought about serious political concerns over the meaning of *national sovereignty.*

The length of the early concessionary contracts is a reflection of the lack of *private property* in land, and the desperate need of the local rulers of these pre-capitalist societies for a steady source of income in order to maintain their stability against the rising tides of the masses. It was in such a circumstance that the international oil companies came to dominate the economy of the entire region, and oil extraction had a lasting imprint on the socio-economic structure of the countries involved. Therefore, it should surprise no one that the duration of the concessions was so long, often extending over three-quarters of a century. D'Arcy's concession, for instance, covered the time span of 60 years. The concession obtained by Kuwait Oil Company was originally 75 years and was subsequently extended to 92 years. The combination of the size and duration of the early oil concessions is, in fact, the outward appearance of the type of political hegemony that international oil companies once exerted upon the pre-capitalist economies of the Middle East during the early part of this century. As we will explain later on, such a political domination, in conjunction with the underlying economic forces, has gradually led to the integration of the above societies into the world economy.

Other characteristics of the early oil concessions are their uniformity, simplicity, and moderate financial terms. The uniformity of terms associated with such concessionary contracts is the result of the monopolization and political domination of a small number of oil companies in their successful attempt to impose a set of general terms on all territories. While it is true that the terms of the early oil concessions were fairly simple, such a simplicity did not reflect any clarity in negotiations. At times the implications of the surrender of oil property rights often were not comprehended by the ruling authorities of the countries involved. It was because of the prevailing pre-capitalist social relations and perhaps the political and social perceptions compatible with them that such contracts were in existence.

Likewise, the existence of 'moderate' financial terms associated with the majority of the early concessions does not have any direct relation to the price of oil during the early period. The royalty rates of the early period were exclusively geared to the extent of political domination of international oil companies and their governments with respect to this particular oil region. Since the economies of these societies in general, and their realm of oil production in particular, were not yet integrated into the global economy, there was no objective mechanism for joint determination of oil prices and oil royalty rates at the international level during this period. It took nearly fifty more years for the oil production to turn into a full-fledged modern capitalist industry, with adequate institutional as well as social mechanisms in order to produce unified values, market prices, and economic rents (royalties plus other sources of oil revenue).

Thus, the financial obligation of the oil companies to the ruling authorities of the Middle Eastern countries during the early period represents the primitive form of economic royalty and the genesis of oil rent of today. The origin of this primitive form, as we argued earlier, was the existence of *old* property relations in conjunction with the political climate within which the surrender of oil property rights had taken place. The early royalties were fixed and usually calculated by the amount of crude oil produced and exported. Even though the term profit-sharing was included in some early concessions, its actual implementation was never realized. For instance, the Anglo-Persian concession, in addition to the prevailing fixed royalty, contained a profit-sharing provision of 16 per cent. However, the framework for the determination and assessment of company profits was always in dispute.

3. From 50–50 Profit-Sharing to Equity Participation: the Era of Transition

During the 1950s the international oil industry witnessed a series of new developments which eventually led to the formation of the Organization of Petroleum Exporting Countries (OPEC) in 1960. The nationalization of the Mexican oil industry in 1948, which was followed by the Iranian oil nationalization of 1951, are both remarkable examples of these developments that reflected the serious concern for political and economic sovereignty which could be found in most oil-producing countries. At the same time, the production of oil gained tremendous momentum during the years following World War II, and a modern oil industry started to develop in the entire region.

From the economic viewpoint, as the oil industry developed further during this period, the conditions for the surrender of oil property rights, i.e., the fundamental basis of oil exploration, became much more complex. It is true that during the above period the oil industry had not yet developed into a full-fledged capitalist sector. For instance, there was no adequate mechanism for the objective formation of values, prices, and oil royalties at the international level. But the signs of a gradual transformation of the relations of production in the industry were readily present in the contract negotiations.

The regime of *50–50 profit-sharing* is the product of this transition. This scheme called for equal division of profits between the oil exporting governments and international companies.[6] In the early period, a fixed royalty rate was the prevailing mechanism. But as the modern oil industry developed, the material basis for the economic foundation of capitalism gradually came into existence. The usage of 'posted price', i.e., a *price* that is devised for long-term oil transactions and contracts, and its connection with the determination of oil royalties is a *significant* development of this period. The scheme of 50–50 profit-sharing is also the outcome of a set of objective conditions that signalled the transformation of existing conditions in the international oil industry of the time.

From the political viewpoint, there was a serious attempt by the oil companies and their respective governments to avoid and extinguish the threatening fire of nationalization. On the other hand, there was little hesitation to go ahead with the new scheme of profit-sharing. As a result there was hardly any resistance by the international oil companies against 50–50 profit-sharing, as long as there existed a framework for the determination of company profits. Of course, since the companies did not wish to expose their *real profit picture* by disclosing their detailed internal documents, they applied the notion of 'posted price' in order to solve the problem.[7] The 'posted price' is a measure which was primarily used by oil companies to value crude oil as it was transferred from one subsidiary to another within their own international network.

Another aspect of this transitional period is the reaction of the oil industry to the nationalization of the oil industry in Iran. Together with their respective governments, international oil companies initiated an economic blockade and collectively boycotted Iranian oil for the period of two years. Their boycott finally ended with the U.S. sponsored coup d'etat against the popular government of Dr. Mossadegh, that brought the Shah back and put the Iranian masses under his despotic rule for some twenty seven more years. This led to massive bloodshed and the eventual *de-nationalization* of the Iranian oil industry.[8] An interesting point is that the companies were always willing to compromise on many issues, except the question of national sovereignty.

The most significant event after the denationalization of the Iranian oil industry in 1954 is the formation of OPEC by five principal oil exporters.[9] The primary objective of these principal oil exporters was to stabilize their source of income. During the 1950s, companies had unilaterally reduced the 'posted prices'. That in turn resulted in the reduction in the income of oil exporting governments. As a result, OPEC was founded in order to organize the effort of the principal oil exporting countries to challenge the oil companies in a collective manner. However, as will be discussed in the next section, OPEC did not prove to be very effective until the early 1970s.

The reason that the period of 1950–70 is called the transition period is the fact that the Middle Eastern oil industry moved very rapidly toward becoming integrated into the world petroleum market. Development of the scheme of 50–50 profit-sharing, along with the designation of 'posted prices' for allocation of royalty and oil rent, is clearly one qualitative step beyond the static mechanism of royalty of the early concessions. In addition, the participation of oil-exporting governments in joint exploration and production with the multinational oil companies was another giant step towards further integration.[10] In some instances oil facilities were owned jointly by companies and oil-exporting governments. For instance, the 50–50 joint-ownership agreement between the 'National Iranian Oil Company' (NIOC) and (Assienda Generale Italiana Petroli) AGIP Mineraria Company for oil exploration and development was formed in early 1957. Similar ownership-sharing schemes gradually spread to the other oil-exporting countries of the region.

Most importantly, because during this period the Middle East oil region produced a significant portion of world oil, the 'posted price' of the early 1950s, which had once been determined unilaterally by the oil companies, became more *responsive* to the material conditions of oil production. As Figure 2 shows, the share of both OPEC and the Middle East oil production has increased substantially during the post-World War II period. At the same time, due to the discovery of large oil fields in the Middle East, the center of gravity of proven world oil reserves has sharply shifted to the Middle East (see Table 4).

This movement toward a more organic relationship between the least productive and most productive regions in the international oil industry was still in progress. During this period, the spot market for a number of oil products came into existence, but the spot market for the most important commodity, i.e., crude oil, did not emerge until the early 1970s. It is true that the 'posted prices' of the early 1950s were somewhat different from those of the late 1960s. But the entire period of 1950–70 is the period of transition from *discretionary determination* of the oil prices to the era of objective value and price formation at the international level.[11] During the period of 1950–70, the conditions of concessions and the nature of contracts contained some elements of the past as well as characteristics of future developments.

4. *Characteristics of Oil Concessions in the Era of Integrated International Oil Production*

The era of early oil concessions was the period of *direct* political domination of the international oil companies over the oil region of the Middle East and elsewhere in the 'Third World'. Therefore, what is known as the oil *royalty* was often unilaterally determined by the oil companies. However, as oil production gained substantial importance and social conditions for the establishment of modern industry came into existence, conflict over the determination of the share of oil-exporting governments has become more acute. In addition, as the world petroleum market expanded and oil - producing regions became parts of one organic whole, new economic as well as political mechanisms were articulated in order to conform to the changing conditions. Thus, the sphere of oil production became further subjected to the laws of value formation rather than arbitrary political coercion. In this connection, a general overview of the history of the early 1970s, with respect to the OPEC declarations and the

Figure 2

World Oil Production Shares, 1940–1975

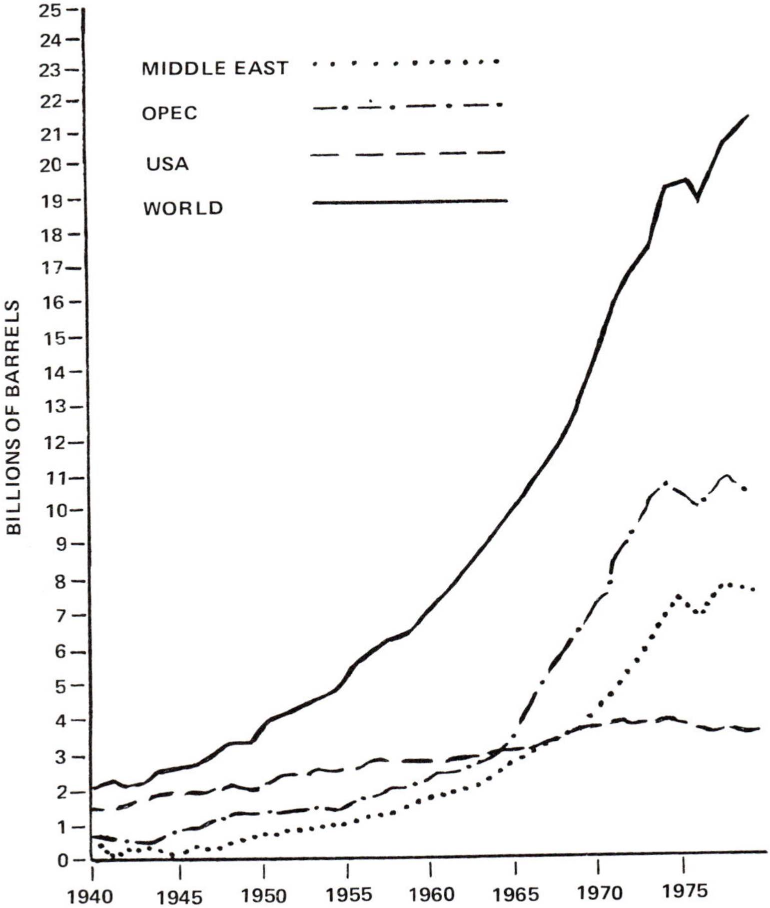

Source: Robert Mabro, *World Energy: Issues and Policies,* ed., (Oxford University Press, 1980), p. 284.

Table 4

Crude Oil Discoveries 1944-74
(billions of barrels)

	World Total	*OPEC*	*Middle East*	*Centrally Planned Countries*	*Capitalist Countries*	*Others*
1974–	115	75	64	18	14	8
1973–1969	51	28	22	11	9	3
1968–1964	38	24	17	8	4	2
1963–1959	19	13	9	2	3	1
1958–1954	34	25	21	4	4	1
1953–1949	16	11	10	1	4	–
1948–1944	7	5	4	–	2	–
		As percentages				
1974–	100	65.2	55.6	15.6	12.2	7.0
1974–1969	100	54.9	43.1	21.6	17.6	5.9
1968–1964	100	63.1	44.7	21.0	10.5	5.4
1963–1959	100	68.4	47.4	10.5	15.8	5.2
1958–1954	100	73.5	61.8	11.8	11.8	2.9
1963–1949	100	68.7	62.5	6.2	25.0	–
1948–1944	100	71.4	57.1	–	28.6	–

Source: Calculated by the author from C. Fliakos and R. Lewison, *The Journal of Energy Development,* Autumn, 1975, pp. 78–79, Tables 4 and 5.

oil companies' responses, is necessary if one is to follow the sequence of events that led to the oil crisis of 1973–74.

In the late 1960s, OPEC issued a series of declarations and resolutions in order to break from the framework of existing oil contracts. In June 1968, the majority of member countries called for participation in the ownership of foreign-owned production facilities (equity participation), the determination of 'posted prices' by the members, and the designation of a mechanism to connect the price of oil with those of other imported industrial commodities from the West, in order to ensure that no deterioration in the price of oil might occur.[12] During the same year, Algeria pressed for the revision of the 1965 Franco-Algerian contract and complained about the slow rate of oil exploration.[13] A year later, Saudi Arabia, while opposing the demand put forth by the less accommodating members of OPEC for nationalization of foreign oil, proposed *equity participation*. In December, 1969, the OPEC members expressed support and sympathy for Algeria and Libya which were at odds with the oil companies.[14] By September, 1970, the Libyan government negotiated a \$0.30/bbl increase, from \$2.23/bbl to \$2.53/bbl and a 5 per cent increase in profit sharing.[15]

Subsequently, the Tehran conference became the site of another conflict between OPEC and the companies. On February 14, 1971, six Persian Gulf oil-exporting countries concluded a five-year agreement with the representatives of 23 international oil companies to have an initial increase of \$0.35/bbl, along with additional \$0.05/bbl increases the following years. Also, for the first time a 2.5 per cent provision for the rate of inflation was incorporated in this agreement.[16]

The Tehran and Tripoli agreements were not the isolated regional incidents that they first appeared to be. Following an increase in the 'posted price' in the Persian Gulf region, Venezuela followed suit. At the same time, the devaluation of the U.S. dollar in August 1971 was the subject of another round of negotiations which finally, on January 20, 1972, resulted in an 8.49 per cent increase in 'posted prices'.[17] Nevertheless, as the U.S. dollar declined further, the need for an upward revision in posted prices, which were for obvious reasons denominated in U.S. currency, became evident. At the Geneva meeting of June, 1973, the question of the automatic adjustment of posted prices to the U.S. dollar was satisfactorily resolved.[18] These negotiations ultimately led to an increase in the level of posted prices in order to compensate for the devaluation of the dollar (see Table 5).

The subject of the negotiations also went beyond 'posted prices'. Concern of OPEC for adjustment of inflation, devaluation of denominating currency, and involvement in equity participation are signs of qualitatively different social relations developing in the international oil industry. OPEC members were already entitled to a portion of crude oil comparable with their shares in equity participation, and such a condition led to a further development of the crude oil *spot market,* which had been virtually non-existent before the 1970s.

The emergence of the crude oil *spot market* is an indication of the internationalization of exchange in the crude oil market. It is a manifestation of the reorganization of production at the global level. The spot market is involved in one-time transactions in crude oil, as opposed to the usual long-term contracts between OPEC and the oil companies. As a result, the spot market prices themselves have had an impact on the terms of the long-term contracts since the time of their development. Prior to the 1970s, there was no spot market in crude oil. The OPEC countries used to sell their share of oil royalty-in-kind directly to the major oil companies at 'buy back' prices. The 'buy back' prices are referred to as prices which were often below the 'posted prices'. However, during the early 1970s, some of the OPEC members were able either to sell their oil directly to the smaller companies or look for the highest bidder at the international market.

Finally, following the October war of 1973, there was a double jump in the posted prices which culminated in the fourfold increase by Arab and non-Arab members of OPEC alike. The summary of posted price increases during the period of 1971-74 is shown in Table 6 and a sample of the resultant increases in the revenue of oil-exporting governments is shown in Table 7. At the same time, Table 8 depicts the relationship between Table 6 and 7, i.e., the calculation of oil revenue per barrel from the 'posted prices'.

Table 5

Posted Prices: Arabian Light 34° Gravity Crude Oil Fob Ras Tanura (dollars per barrel)

December 31, 1960	1.80
December 31, 1965	1.80
December 31, 1970	1.80
February 15, 1971	2.18
June 1, 1971	2.29
January 20, 1972	2.48
January 1, 1973	2.59
April 1, 1973	2.74
June 1, 1973	2.90
July 1, 1973	2.96
August 1, 1973	3.07
October 1, 1973	3.01
October 16, 1973	5.12
November 1, 1973	5.18
December 1, 1973	5.04
January 1, 1974	11.65

Source: Foster Associates, *Energy Prices, 1960-73* (Cambridge, Mass.: Ballinger Publishing Company, 1974), p. 18, Table 3.3.

Table 6

Posted Price Changes of Selected Oil-Exporting Countries, Various Dates, 1971-74
(dollars per barrel)

	Iran (34°)	Saudi Arabia (34°)	Venezuela (35°)	Libya (34°)
January 1971	1.79	1.80	2.344	2.55
February 15, 1971	2.17	2.18	2.799[1]	3.447[2]
January 20, 1972	2.476	2.479	3.261	3.673
January 1, 1973	2.579	2.591	3.3621	3.777
October 16, 1973	5.341	5.119	7.802[3]	8.925
January 1974	11.875	11.651	14.247[4]	15.768

1 Mar. 18 price.
2 Mar. 20 price, following Tripoli agreement.
3 Nov. 1 price.
4 Excluding sulfur premium.

Source: For Venezuela and Libya 1973-74, 'Petroleum Economist' (formerly 'Petroleum Press Service'), various issues; for Iran and Saudi Arabia, 'Middle East Economic Survey', December 28, 1974, p. 5; for Venezuela and Libya 1971-72, U.S. Department of the Interior, Office of Oil and Gas, 'Worldwide Crude Oil Prices, Summer 1972', Table IX.

Table 7

OPEC Oil Revenues: 1971-74
(dollars per barrel)

	Iran (34°)	Saudi Arabia (34°)	Venezuela (35°)	Libya (34°)
January 1971	1.103	1.099	1.62	1.748
February 15, 1971	1.370	1.373	2.023[1]	2.382[2]
January 20, 1972	1.55	1.548	2.315	2.527
January 1, 1973	1.617	1.616	NA	2.590
October 16, 1973	3.292	3.148	5.09	5.89
January 1974	7.253	7.108	8.77	10.04

1 Mar. 18 price.
2 Mar. 20 price.

Source: For Iran and Saudi Arabia, 'Middle East Economic Survey', December 28, 1973, p. 7; for Venezuela 1971 and 1972, U.S. Department of the Interior, Office of Oil and Gas, 'Worldwide Crude Oil Prices, Summer 1972', for 1973 'The Economist', January 5, 1974, p. 63; for Libya 1971, January 1973, U.S. Department of the Interior, Office of Oil and Gas, 'Worldwide Crude Oil Prices, Summer 1973'. For October 1973, and January 1974, 'The Petroleum Economists', February 1974, p. 43.

Table 8

The Calcualtion of Oil Revenue per Barrel of Light Arabian Crude for the OPEC Countries

	Nov. 1965	*Jan. 1973*	*July 1973*	*Jan. 1974*	*Jan. 1975*	*Jan. 1976*	*Jan. 1977*
1. Operating Cost	$ .10	$ .10	$ .10	$ .10	$ .12	$ .12	$ .12
2. Royalty[a]	.22	.32	.38	1.45	2.25	2.47	2.60
3. Subtotal	.32	.42	.48	1.55	2.37	2.59	2.72
4. Posted Price	1.80	2.59	3.01	11.65	11.25	12.38	12.99
5. Less Cost and Royalty (3)	−.32	−.42	− .48	−1.55	− 2.25	− 2.59	− 2.72
6. Subtotal	1.48	2.17	2.53	10.10	9.00	9.79	10.27
7. 'Tax Paid'	.74[b]	1.19[c]	1.40[c]	5.55[c]	7.55[c]	8.31[c]	8.73[c]
8. Govt. 'Take' (7 + 2)	.96	1.51	1.88	7.00	9.80	10.78	11.33

a at 12.5%
b at 50%
c at 55%

Source: John M. Blair, *The Control of Oil* (New York: Pantheon Books, 1976), p. 269 and *International Economic Report of the President* (Washington, D.C.: US Government Printing Office, 1977), p. 175, Table 65(c).

Another major development during the war and embargo period is the bidding called for by the Iranian government for delivery of 470,000 barrels a day of crude oil on December 10, 1973. This was a major test for the crude oil spot market, and the selling price was over $17 a barrel.[19] The same practice by Nigerian and Libyan governments resulted in a $20 to $24 price range toward the end of 1973.[20] Some members of OPEC actually called for the designation of the existing spot prices as posted prices, but Saudi Arabia opposed it and instead they reached a compromise for a posted price of $11.61/bbl (Arabian light 34° API).[21]

To summarize, during the period of the late 1960s and early 1970s, tremendous institutional changes took place within the international oil industry. The constant demand for an increase or adjustment in *posted prices* by OPEC is the most visible aspect of this period. Concern over accommodation for inflation and currency devaluation is another indication of the integration of the Middle Eastern oil industry into the global economy. The phenomena of equity participation and development of crude oil spot market are additional features of this period.

As we stated in Section 3, the period of 1950–70 has the socio-economic features of the past as well as the characteristics of the future. During this period, oil rent gradually developed, took on real form, and ultimately became a full-fledged relation associated with the capitalist mode of production (CMP) as the oil values and market prices conformed to the laws of internationalization. (See Table 9.)

Table 9

*Comparison of the Middle East Posted Prices and the U.S. Wellhead Prices of Crude Oil**
1961-1978

Year	*Posted Price Arabian Light*	*Average U.S. Oil Price*	*Absolute Value of the Percentage Difference***	
1961	1.80	2.89	38%	
1966	1.80	2.88	37%	1961-72
1970	1.80	3.18	43%	average:
1971	2.24	3.39	34%	35.7%
1972	2.48	3.39	27%	
1973	3.97	3.89	2%	
1974	11.38	10.13	12%	1973-78
1975	11.81	12.03	1.8%	average:
1976	11.69	12.16	3.8%	2.7%
1977	12.39	13.59	8.8%	
1978	12.70	13.95	9%	

* Figures in U.S. dollars.
** U.S. price is the base.

Source: Compiled by the author from API, *Basic Petroleum Data Book,* 1981, Section VI, Table 10; DOE, *Monthly Energy Review,* various issues.

NOTES

1. Henry Cattan, *The Evolution of Oil Concessions in the Middle East and North Africa* (New York: Oceana Publishers, Inc., 1967), pp. ix-x.
2. See John Blair, *The Control of Oil* (New York: Pantheon Books, 1976), pp. 153-155; and P.S. Dasgupta and G.M. Heal, *Economic Theory and Exhaustible Resources* (Cambridge: Cambridge University Press, 1979), p. 50.
3. Henry Cattan, p. 2.
4. *Ibid.,* p. 4. See also Zuhayr Mikdashi, *A Financial Analysis of Middle East Oil Concessions, 1901-65* (New York: Praeger Publishers, 1966).
5. See V.I. Lenin, *Imperialism, The Highest Stage of Capitalism* (New York: International Publishers, 1970), Chapter V.
6. Henry Cattan, *The Law of Oil Concessions in the Middle East and North Africa* (Dobbs Ferry, New York: Oceana Publications, Inc., 1967), Chapter 5.
7. As we stated earlier, during the era of early concessions in which some sort of profit-sharing was present, companies and ruling authorities were always in dispute.
8. The concrete account of this can be found in 93rd Cong., 2nd Session, U.S. Senate *Hearing on Multinational Petroleum Corporations and Foreign Policy,* 1974, Part 7.
9. These countries are Iran, Iraq, Kuwait, Saudi Arabia, and Venezuela. See Fuad

Rouhani, *A History of OPEC* (New York: Praeger Publishers, 1971); and Zuhayr Mikdashi, *The Community of Oil Exporting Countries* (Ithaca, New York: Cornell University Press, 1972), for additional information.

10. Henry Cattan, *The Evolution of Oil Concessions in the Middle East and North Africa* (Dobbs Ferry, New York: Oceana Publications, Inc., 1967), Chapter 4, pp. 122-51, and Appendix 1.
11. It is worth mentioning that the OPEC 'posted price' of oil, after the oil crisis of 1973-74, is not very much different from the price of North Sea oil, for the long-term contracts. However, long before the crisis, the OPEC posted prices were often the subject of arbitrary determination of oil companies and the oil-exporting governments.
12. Edith Penrose, 'The Development of Crisis', *DAEDALUS, Journal of the American Academy for Arts and Sciences,* Fall 1975, p. 40. See also Edith Penrose, 'Government Partnership in the Major Concessions', *Middle East Economic Survey,* August 30, 1968.
13. *Ibid.*
14. *Ibid.*
15. The scheme of 'profit sharing', which has been implemented since the 1950s is often referred to as an income tax levied by the local governments. We will explain in the section on the theory of oil rent that such a reference is erroneous.
16. Ad Hoc Committee on the Domestic and International Monetary Effect of Energy and Other Natural Resource Pricing (Committee on Banking, Currency and Housing), 94th Congress, First Session, *The Economics of Energy and Natural Resources Pricing,* March 1975, p. 89.
17. *Ibid.*, p. 89.
18. *Ibid.*, pp. 89-90.
19. Edith Penrose, 'The Development of Crisis', *DAEDALUS,* Fall 1975, p. 51.
20. *Ibid.*
21. *Ibid.*

PART TWO

THEORETICAL DEVELOPMENT AND EMPIRICAL ANALYSIS OF THE OIL CRISIS

CHAPTER IV

THE ROLE OF RAW MATERIALS IN THE PRODUCTION OF VALUE

1. Fluctuation in the Price of Raw Materials

The oil crisis of 1973–74 was in fact a distinct case of fluctuations in the price of raw materials. The impact of such a drastic change in the price of crude oil has, in turn, produced a change in the price of auxiliary materials used in many industries and has, therefore, exerted a substantial influence on the value of *constant capital* used in the process of production of a great number of commodities. In this connection, Marx maintained that

> So far as machinery is concerned under this head (i.e., raw materials), its own raw material consists of iron, wood, leather, etc. Its own price is therefore affected by fluctuations, either in the price of the raw materials of which it consists, or of the auxiliary materials consumed in its operation, the rate of profit falls *pro tanto.* And vice versa.[1]

Crude oil and most of its derivatives or by-products which are used as *auxiliary materials* in the production process are part of the more fundamental category of *constant capital.* Therefore, it would be evident that fluctuations in the price of crude oil tend to influence the magnitude of the price of production of the elements of *constant capital* and, therefore, the general rate of profit in the economy. In this manner, theoretically speaking, the fourfold increase in 'posted price' of crude oil, which was the reflection of the reorganisation of the oil industry at the international level, must have resulted in the fall of the general rate of profit, *ceteris paribus.*

Our conclusion holds true even if the rate of exploitation as well as the mass of surplus value in the labor process remain constant. The raw materials, or more specifically auxiliary materials, are the circulating component of the constant capital in the production process. Since [Surplus value (s)/constant capital (c) + variable capital (v)] constitutes the formula for the rate of profit, then when C's value increases, *ceteris paribus,* the rate of profit will obviously decline.[2]

We will prove in the following chapters that in fact the oil price increase of the 1973–74 period was the consequence of the increase in the magnitude of the newly established market value within the international oil industry. This increase has also necessarily led to an increase in the value of constant capital and decrease in the general rate of profit through this particular crisis.

It is important, at the outset, to note that the price in the above sense refers to the *price of production* as opposed to the *market price*. The former is when the production of value includes the transformation problem. This price is the center of gravity of market prices. Nevertheless, there are a large number of contributing factors, such as the variation in supply and demand, by virtue of which changes in the magnitude of the market prices take one direction or the other. The price of production is the transformed form of value. This transformation is due to the existence of different composition of capital associated with the various sectors of the economy. The market prices are the ones which fluctuate around the prices of production, given the state of supply and demand. In this manner, the market price is the phenomenal form of the price of production, and it is not necessarily identical with it. If there is an excess supply for a particular commodity in a market, the market price of that commodity drops below the price of production. If, on the other hand, there is a condition of excess demand, it leads to market prices which are above the prices of production. Since the equalization of demand and supply in market economy is accidental, the market prices tend to regularly fluctuate around the prices of production.[3]

2. *Importance of Raw Materials: An Examination of the Distinction Between Exhaustible and Renewable Resources*

The importance of raw materials, in general, and that of crude oil in particular, has been emphasized by the majority of present-day economists. However, the majority of writers on the subject of the oil crisis have exclusively emphasized the physical characteristics of oil in the production of commodities. For instance, the usage of this common approach is clearly visible in the new studies in conjunction with the *tri-factoral production functions,* with the inclusion of natural resources as the third factor.[4]

Nevertheless, regardless of the conceptual and empirical problems associated with these production functions, the very inclusion of natural resources as an additional factor is the reflection of confusion between use value and exchange value in neoclassical economics. As a result, there is a common emphasis these days to distinguish *exhaustible* from *renewable* resources. It is argued that since the rate of renewability of exhaustible resources is zero, such resources, if utilized, will eventually be exhausted in the future. On the other hand, the renewable resources tend to grow or shrink over time, depending upon their rate of utilization and renewal. It is further stated that the so-called exhaustible resources, while *commodities,* are *non-produced commodities.* This means that since exhaustible resources cannot be reproduced in the same manner as other commodities, including the commodity *labor power,* and given the fact that the subject matter of political economy is the analysis of the produced commodities, then, the category of *value* is inadequate for the analysis of production and distribution of exhaustible natural resources.[5]

It is clear that raw materials, both in their *auxiliary* and *direct* forms, are indispensable for the production of value. Raw materials, as part of nature, have been also indispensable to production in general throughout history, under all the previous modes of production. Consequently, insofar as the physical attribution of resources in question is concerned, as use values they have been indispensable to the production process in both capitalist as well as pre-capitalist modes of production. Nevertheless, if there is any distinction at all between capitalist social relations and those of pre-capitalist modes, it is not so much for the fact that natural resources are *physically indispensable,* but rather that they are *socially indispensable* for the production of value. Connected with the latter category, these resources are *materials of labor* in the production of surplus value. Therefore, if one holds that the subject of modern political economy is not the study of undifferentiated social relations in human history, exclusive emphasis on the category of use value will undoubtedly distort one's perspective. If in capitalism, the existing social relations take the form of the relation of *things,* production of natural resources is also part and parcel of this labor process. Now if there exists any distinction between capitalist and pre-capitalist modes of production, it obviously is *not* due to the production of *use-value,* for production of *use-value* is common to all modes of production. What is not common to all is indeed the production of *exchange-value,* that is particular to production under capitalism alone. As a result, the division of resources as materials of labor into the fictitious categories of exhaustible and renewable resources is not fruitful for understanding of the laws of motion of a particular resource utilization under capitalism. Also the mere exhaustibility of resources in its physical sense does not seem to be a *barrier* to the production and reproduction of capital. As the long history of capitalism shows, there are a great number of occasions when particular resources and raw materials were utilized at one time, and later on either physically *exhausted* by themselves or *abandoned* due to economic reasons. But capitalism continued to substitute and utilize other types of natural resources.

Likewise, there are many resources which are presently being used in the process of production which may be exhausted in a shorter or longer time in the future. Nevertheless, value production will not face any barrier for its self-expansion. *For the barrier of capital is capital itself* and not the natural objects utilized by it. Therefore, the importance of raw materials, and, for that matter, natural resources, does not so much associate with their physical characteristics, but rather relates to the barrier of the existing property relations that may be involved in their production. It is in this connection that the phenomenon of landed property in agriculture or the specific property relation associated with the oil industry (which will be discussed in forthcoming chapters) is so significant.

3. Importance of Raw Materials: Marx's View

The utilization of raw materials is closely related to the development of labor productivity that would lead to an increase in the mass and the value of constant capital. This increase in mass, due to constant technological development, is proportionally larger than an increase in the corresponding values. Thus, the industries that technically tend to consume a significant mass of raw materials with an increase in labor productivity, tend to require even larger masses of raw materials. Nevertheless, depending upon the proportionate change in the labor productivity associated with the raw material-producing sectors, the value of raw materials will change. If, in fact, the labor productivity of the raw material-producing sectors does not increase in the same proportion as the labor productivity in raw material-consuming spheres, the values of raw materials tend to increase relative to those of the finished commodities.

The above analysis, at the same time, gives us a clear perspective in recognition of the concept of *constant capital* as opposed to *fixed capital.* The former category, which includes the material of labor, both in direct and auxiliary form, is often confused with the latter by most economists of today.

4. Conclusion

To sum up, the raw materials are part and parcel of the production of value. The auxiliary raw materials are used in their entirety in a period of production just as the material of labor. With the consumption of these materials in the production process, their value is preserved as they become transformed into the finished commodities.

However, if the value of the above materials changes, it would necessarily lead to a change in the value of constant capital. Given a particular rate of exploitation in the labor process, a change in the value of constant capital in the production process will, in turn, lead to a change in the rate of profit through the S/C + V formula. Also, an increase in the value (or price of production) of constant capital induced by an increase in the production price of the raw materials does not necessarily lead to a decline in the rate of profit through crisis.

If the 1973–74 increase in the price of crude oil is the result of the reorganization of production at the global level and formation of newly established social value of larger magnitude, then one has to expect that the crisis that resulted from such a reorganization will also be associated with a decline in the rate of profit. In the following chapters we shall maintain that, in fact, the market value of crude oil has changed in conjunction with the oil crisis. The above relationship also demonstrates that the emergence of an economic crisis in one sector can possibly lead to a wider economic crisis of general magnitude. Finally, as we have argued, the distinction between renewable and exhaustible raw materials is a superficial one, as it relates to the realm of use value alone. From the viewpoint of value production, the physical characteristics of raw materials are nothing more than *phenomenal forms* through which the existing social relations are manifested and reproduced in the society.

NOTES

1. Karl Marx, *Capital,* vol. III (New York: International Publishers, 1967), p. 106.
2. As Marx put it, '. . . like those variations which result from economy in the use of constant capital, variations resulting from fluctuations in the price of raw material also *always* affect the rate of profit, even if they leave wages, and thus the rate and mass of surplus-value, completely undisturbed'. Karl Marx, *Capital,* vol. 3 (New York: Vintage Books, 1981), p. 200, (emphasis added).
3. Karl Marx, *Capital,* vol. 3 (New York: Vintage Books, 1981), p. 279.
4. For further observation, see the papers presented in the Symposium on Exhaustible Resources held in London that were subsequently published in a special issue of the *Review of Economic Studies,* 1974.
5. See for example, P.C. Thanh, 'Growth and Optimal Growth with Natural Resources', February 1977, Department of Economics, The American University, Washington, D.C. (Mimeographed).

CHAPTER V

THEORIES OF RENT AND THEIR CRITIQUE: THE DEVELOPMENT OF A THEORY OF OIL RENT

The purpose of the present chapter is to develop a theory of rent for oil production in order to capture the historical significance of the structure of property relations in conjunction with the production conditions in the oil industry. However, prior to the development of a specific theory of rent for the oil industry, it is appropriate to study the specificity of the notion of rent in capitalism. This of course leads us to a brief evaluation of a selected number of rent theories which are historically associated with the analyses of agricultural production.

1. Ricardo's Notion of Rent

'If all land had the same properties, if it were unlimited in quantity, and uniform in quality,' Ricardo pointed out, '*no charge could be made for its use*, unless where it possessed peculiar advantages of situation'.[1]

Clearly, both limited quantity and heterogeneity of land are factors that led Ricardo to develop his version of rent theory in agriculture. He argued that as the margin of cultivation is extended to use land of inferior quality, cultivation on the land of higher quality results in rent. The rent of this type is called differential rent by Ricardo. Moreover, this differential rent reflects the difference in the quality of 'marginal' and that of 'intra-marginal' land as cultivation is extended. Another way to perceive such a rent is by noting the limited quantity or *scarcity* of land. That is why Marshall, in evaluating Ricardo's conception of rent, pointed out that, 'In a sense, all rents are scarcity rents, and all rents are differential rents.'[2] The origin of rent, as Ricardo sees it, is due to the physical characteristic of nature, which is perceived to be universal no matter what the *nature* of production might be. As a result, this concept is applicable to all periods of history. In fact, the consequence of such an analysis is that the phenomenon of rent is *not* unique with capitalism.

Ricardo attempted to make a distinction between rent and profit as a part of a more general theoretical development, i.e., the laws which regulate the process of distribution of 'the produce of the earth' among the 'three classes of community, namely, the proprietor of the land, the owner of the stock or capital necessary for its cultivation, and the laborers by whose industry it is cultivated'.[3] It is not surprising at all, therefore, that the *origin* of rent, in Ricardo's sense, evolves through the *technical consideration* rather than *social necessity*. The cause of rent here is seen in the extension of cultivation instead of the monopolization of nature which necessarily leads to the development of a particular structure of property relations in agriculture.

2. *Marshall's Conceptual Framework*

Since the Marshallian theory of rent is halfway between the Ricardian and neoclassical conceptions of rent, reference to his theory is necessary. 'While man has no power of creating matter,' Marshall said, 'he creates utilities by putting things into useful form; and the utilities made by him can be increased in supply if there is an increased demand for them: they have a supply price.'[4] Marshall continued to say that although land is among the objects of utilities, one cannot, simply, exercise any control over it due to the fixity of the supply of nature. Therefore, land has no particular price.[5] Marshall did make a clear distinction between land and the products of land, when he stated that 'the fundamental attribute of land is its *extension*.'[6]

It should be noted that the similarity of the Marshallian conception of rent and its counterpart in the Ricardian tradition is not only due to the emphasis on the extension of cultivation. Marshall, like Ricardo, considered the 'inherent properties, which the land derives from nature. . .' and recognized that 'It is chiefly from them that the ownership of agricultural land derives its peculiar significance, and the Theory of Rent its special character'.[7]

However, there are striking dissimilarities between the Marshallian and Ricardian theories of rent: Ricardo attempted to make a distinction between *rent* and *profit*, while Marshall tried to do exactly the opposite. Marshall remarked that:

> . . . the full rent of a farm in an old country is made up of three elements: the first being due to the value of the soil as it was made by nature; *the second to improvements made in it by man;* and the third, which is often the most important of all, to the growth of a dense and rich population, and to facilities of communication by public roads, railroads, etc.[8]

It seems evident from the above passage that, first of all, the concept of rent is caused by nature. Second, rent is a universal concept and therefore does not correspond to a specific period of history. Third, the concept of private property and monopolization of nature by the owners of land is taken for granted. Fourth, the growth of mass communication and transportation as the source of one form of rent is heedlessly mixed up with the ground rent. Fifth, not only is there no clear distinction between rent and profit as can be found in the works of Ricardo and Marx, but also Marshall has often misunderstood the implication of the theoretical contributions of his predecessors.

Despite these differences, the concept of rent formulated by Marshall gives rise to a concept of rent which is not modified by the characteristics of different historical epochs. Rent is a universal concept in Marshall's theory as well as in Ricardo's.

3. Marx's Approach: Historical Specificity of Capitalism

The formulation of Marx's theory of rent in agriculture is a direct consequence of his critique of political economy. He argues that the concept of capitalist ground rent is rather based on the social conditions of the development of capital in agriculture. As a result, a modern notion of rent has to be perceived within the capitalist mode of production alone, not as a universal phenomenon that is relevant to all history.[9]

The basis of capitalist rent in Marx's sense lies in the production of *value*, in contrast with the rent in the Middle Ages which is connected to the physical entity of the product and its *use-value*. Thus, the notion of capitalist rent as opposed to rent in pre-capitalist production does not have its origin in nature.

As Marx pointed out,

> An erroneous concept of the nature of rent is based upon the fact that rent in kind. . . has been dragged over into modern times from the natural economy of the middle ages, completely in contradiction to the conditions of capitalist mode of production. It thereby creates the impression that rent does not arise from the price of the agricultural products, but from its mass, thus not from social conditions, but from the earth.[10]

In order to discover the genesis of capitalist ground rent, one needs to look into the relations of use-value and the exchange-value in production of commodities. It is quite true that without the production of use-value there is no possibility of producing exchange-value. But in order to be able to make a *distinction* between the capitalist mode of production, on the one hand, and all the other pre-capitalist modes of production, on the other hand, one needs to look for differences and not the similarities. It is only under capitalist social relations of production that the production of *exchange-value* is based on the exchange of the commodity *labor power.* In this connection, commodities have two different aspects: (1) value in use, and (2) value in exchange. A pint of milk as a commodity and a very similar pint of milk produced under pre-capitalist social relations, cannot be distinguished from the appearance if one does not know in advance under what social relations each one is produced. Likewise, if the conditions of production under which these two pints of milk were produced are totally *abstracted* once they are produced, no one can tell the difference.

Thus, in order to be able to make a clear distinction between a specific period of history, i.e., capitalism, and history in general, one has to abstract the similarities and concentrate on the differences. While the utilization of nature is common to all humans in history, the act of the utilization of nature in the capitalist production, as opposed to all the previous modes of production, is not an end in itself. It is rather a medium through which the social relations among people are transformed regularly into relations among things.

By the same token, the significance of discovering the genesis of *capitalist rent* is not due to the physical attributes of nature, for these physical aspects by themselves do not lead to the reality of rent. Rather, what gives significance to this notion is the constitution of those social relations of production which include *nature* in value production. The monopolization of a natural sphere, on the one hand, and the expansion of capital into it, on the other, leads to the distribution of surplus-value (rent and profit) before the formation of a general rate of profit. The above processes, i.e., both the monopolization of nature by the landed property, (or any other type of property relation which monopolizes the condition of labor), and the expansion of capital, do not have their origins in the physical characteristics of nature, since these processes are rather the results of the existing social relations of production themselves.

Historically, during the transition from the feudal mode of production (natural economy) to the capitalist mode of production, a dual transformation of (1) the separation of the peasantry from the land and the transformation of this class into the class of wage laborer, and (2) the transformation of land into the monopolized private property of the landlords, developed. The transformation of nature into private property is initiated by capital itself which, in turn, comes to stand against it in the form of rent. Thus, capitalist rent is itself a direct consequence of the rule of capital, and its origin does not have anything to do with the physical characteristics of nature. The notion of rent under the capitalist mode of production is the *phenomenal form* of these social relations which are constantly emerging through the material form of nature. The reflection of the above simultaneous transformations upon the new system is ably detected by Marx, where he states that

> The determination of the market-value of products, including, therefore, agricultural products, is a social act, albeit a socially unconscious and unintentional one. *It is based upon the exchange-value of the product,* not upon the soil and the difference in its fertility.[11]

4. *Theory of Rent; Exhaustible and Renewable Resources*

The genesis of capitalist rent created the social conditions under which a capitalist *rentier* class came into existence. This process as we have seen in the previous section gave rise to the phenomenon that Marx called capitalist ground rent. We have also realized that the genesis of capitalist rent is the direct consequence of the private appropriation or the monopolization of nature, which is specific to the capitalist mode of production. We propose that such an understanding is a pre-condition for any analysis that deals with the utilization of nature and in this manner starting with nature in its particular *form,* i.e., *exhaustible* or *renewable,* is not fruitful.[12]

The genesis of *rent,* according to Marx, as opposed to Ricardo, is not the quality of intra-marginal land, but the form of the manifestation of a specific social relation. The quality of 'marginal land', in association with the concept of zero rent advanced by Ricardo, does not have any relation to the origin of rent, because it tends to exclude the notion of rent as a social relation of landed property. Without being able to explain the origin of rent, Ricardo has developed a special case of differential rent by an arbitrary assumption of moving from better to worse land. This notion of differential rent as a category is devoid of social significance since it is not grounded in production conditions as a social category unique to the capitalist mode of production.

We have argued that in order to analyze the notion of *rent* one needs to unveil its genesis. This investigation of the genesis of rent reveals the importance of the emergence of capitalist social relations. Rent is a *phenomenal* manifestation of the interaction of the structure of accumulation and the structure of landed property in capitalism. Rent is therefore (1) a social category, and (2) an historically unique category and specific to the capitalist mode of production.

We can make a similar criticism of the modern literature's attempt to divide 'exhaustible' from 'renewable' resources. This approach neglects considering the origin of rent; the analysis is *ahistorical.* Under the capitalist mode of production, capital is a self-expanded value. It expands into every facet of life, including the realm of nature. An element of nature, such as an oil pool, which did not have an exchange value in the previous historical epochs, has now been transformed as a necessary condition for the expansion of capital. Therefore, the significance of nature and, for that matter, *natural resources,* is not merely in their use-value. Instead their importance is due to their utilization for the production of surplus value.

5. *Marx's Theory of Differential Rent: Ricardo's Critique*

As we have pointed out, the notion of rent from Ricardo's viewpoint relates to existing differential qualities of land in agricultural production. According to Ricardo, the *order of utilization* starts from the best land and gradually extends to the worst land as the demand for the product of land increases. Since the price of corn is determined by the conditions of production associated with the worst land of zero rent, the production from better land will result in surplus over and above normal profit. That is the reason why Ricardo argues that 'corn is not high because a rent is paid, but a rent is paid because corn is high. . .'[13]

Accordingly, the above conception indicates that as the demand for agricultural products increases and the quality of land under cultivation deteriorates, the price of the production of corn will necessarily be increased. Such an increase will also accompany a higher and higher amount of rent over time.

Even though Ricardo correctly recognized the notion of differential rent in agriculture at the level of distribution, he essentially missed its significance in the production process and its relation to accumulation by accepting the Malthusian theory of population and the inevitable scarcity of agricultural products and articles of consumption.[14] In fact, the key to the conceptual presumption of the simultaneous increase in prices and rents is the assumption of relative scarcity in conjunction with the extension of cultivation from better to worst lands. However, contrary to Ricardo's presumption, the development of the productive forces in agriculture is such that neither the best land nor the worst will always remain the same. An *a priori* assumption of the extension of cultivation from better to worst lands is not only false conceptually, but also incorrect empirically.

In contrast to Ricardo, Marx approaches the phenomenon of rent by rejecting Ricardo's arbitrary assumptions. The precondition for such a rejection was perhaps adequate study of the *actual* direction of movement in the price of corn and the rent of land under cultivation during the first part of the nineteenth century in Britain. In his letter of January 7, 1851, Marx writes to Engels,

> I am writing to you today in order to lay a little question of theory before you. . .
>
> 1. There is no doubt that as civilization progresses poorer and poorer kinds of land are brought under cultivation. But there is also no doubt that, as a result of the progress of science and industry, these poorer types of land are relatively good in comparison with the former good types.
> 2. Since 1815 the price of corn dropped. . . Rent [nevertheless] has continually risen.
> 3. In every country we find, as Petty already noticed, that when the price of corn drops the total rental of the country rises.[15]

The recognition of the above historical facts in conjunction with adequate knowledge about the work of his predecessors are among the factors that contributed to Marx's understanding of the subject, a distinction between (1) the *equal* applications of capital upon the land of *unequal* quality, and (2) the *unequal* applications of capital upon the same land.

In addition to these findings in Volume III of *Capital,* the above distinction can be clearly seen in *Theories of Surplus Value:*

> *With a given capital investment,* the variation in the amount of rent is only to be explained by the varying fertility of the land. The variation in the amount of rent, *given equal fertility,* can only be explained by the varying amount of capital invested.[16]

The above conditions are referred to as differential rent of the first type (DRI) and differential rent of the second type (DRII). It should be apparent that in the actual situation these two types of differential rents are formed jointly in the presence of each other. In other words, the formation of social value is the result of the *unequal* applications of capital to *unequal* qualities of land through competition. Moreover, DRI and DRII are not actually formed independently of each other. The production conditions which lead to the change in the level of one form of differential rent do not necessarily leave the other form unchanged. Ben Fine summarizes the above problem in the following manner:

> For DRI, there is the problem of determining the worst land in the presence of unequal application of capital (DRII). . . for DRII, there is the problem of determining normal level of investment in the presence of differing lands (DRI). . . These problems concern the simultaneous determination of worst land and normal capital in agriculture.[17]

In order to solve this problem, Marx analytically examines the formation of DRII in conjunction with DRI under the changing prices of production. Moreover, instead of attempting to identify the simultaneous determination of *worst land* and *normal capital* directly, Marx let the prices of production decline, remain constant, and increase in order to capture the varying impact of capital investments. In other words, Marx successfully goes beyond the static distributional question of rent in order to address the question of the dynamics of capital accumulation in conjunction with the barrier of landed property in agriculture.[18] At the same time, unlike Ricardo's, Marx's theory of differential rent is quite consistent with his theory of value, which is based on socially necessary labor time, as opposed to various labor-embodied theories.

Another aspect of Marx's critique of Ricardo's rent theory is the rejection of Ricardo's notion of differential rent based upon the margin of cultivation along the 'descending line'. Marx has shown that depending upon the conditions of production there will be movements toward better or worse lands. After making extensive examples, he remarked that, 'In reality, the ascending and descending lines will cut across one another, the additional demand will sometimes be supplied by going over to more, sometimes to less fertile type of land, mine or natural agent.'[19] On the contrary, Ricardo believed that the market value of agricultural commodities is *always* determined by the value produced on the marginal land.

The difference between Marx and Ricardo is great. Marx argues that,

> As soon as the additional supply surpasses the capacity of the market, as determined by the old market-value, each class naturally seeks to force the whole of its product *onto* the market to the exclusion of the product of the other classes. This can only be brought about through a fall in price, and moreover a fall to the level where the market can absorb *all* products. If this reduction in price is so great that the classes I, II, etc., have to sell *below* their costs of production, they naturally have to withdraw [their capital from production] . . . .But it is further clear that in these circumstances it is not the worst land, I and II, but the best, III and IV, which determines the market-value. . .[20]

What Marx states is clearly different from that of Ricardo. The competitive struggle of capital invested on different 'classes' of land and the existence of specific market conditions can certainly lead to formation of a market value different from the value of the product of marginal land. As a result, such a market value regulates the market, and it can be below the value of the marginal producer.

It is necessary at this juncture to add a few words about Marx's theory of *absolute rent.* The notion of absolute rent in Marx is the effect of the domination of landed property in agriculture. As a result, even the worst land normally tends to appropriate a positive rather than a zero rent. The magnitude of absolute rent, however, is determined by the extent to which the successive investment of capital upon the existing land in use can be obstructed by DRII.[21]

It is also important to note that the assumption of zero rent for 'marginal' land by Marx, in the analysis of differential rent, is just used as an illustration of the concept of differential rent and in no way contradicts his notion of absolute rent. In his discussion of absolute rent, Marx further emphasizes that, 'no matter what the case may be as regards the rent of the least fertile soil, the law of differential rent is not only independent of it, but that may be the only manner of *grasping* differential rent in keeping with its character is to let the rent on Soil A equal zero. Whether this actually equals zero or greater than zero is immaterial so far as the differential rent is concerned and, in fact, does not come to consideration'.[22]

To illustrate the mechanism of differential rent, Marx assumes that the 'marginal land' has a zero rent. However, he insists that such an assumption of zero rent is not necessary. Furthermore, contrary to Ricardo, Marx notes that under the capitalist mode of production the formation of rent results from the development of landed property in agriculture, and it is idealistic to argue that rent should not exist. Ricardo did not attempt to analyze the phenomenon of rent based on the internal development of capitalist social relations. Instead, he tried to explain that rent is the result of the difference in fertility and other physical attributes of land, quite independent of landed property. Thus, the subject of rent from his viewpoint relates to *natural* and technical dimensions rather than *social* conditions under which value is produced. Consequently, it is hardly surprising that Richardo's theory of rent is not only *universal* in character, but also eternal as long as there is scarcity.[23] Contrary to this, we have already discussed that the precondition for the formation of differential rent the existence of differential productivity and the resultant differential profitability through competition. However, it is due to the impact of landed property that the differential profitability becomes *permanent* in agriculture.

6. *Theory of Differential Oil Rent*

In this section, we intend to summarize our previous theoretical propositions in order to develop a theory of differential rent for the oil industry. The phenomenon of economic rent as a category distinguished from normal profits is neither original with Marx nor specific to Classical political economy. However, what has made Marx's theory of agricultural rent different from his predecessors' theories is 'the *specificity* of the analysis itself, not the category'.[24]

The notion of oil rent in the oil industry is nothing but the phenomenal form of the specific property relation that is unique to the oil industry. Historically, the *separation* of the ownership of hydrocarbons from the ownership of the oil fields resulted in the development of this barrier within the context of capital accumulation in the oil production. In countries or regions where the ownership of the surface soil legally include the subsoil, the owners of oil, i.e., capitalist producers, are faced with the obstacle of the ownership of the oil lands. This relationship remains the same even if the state as a legal form of landed property owns the oil lands, (since the capitalist mode of production still dominates).[25]

The separation of ownership is part of an historical process which is realized legally through the act of lease contracts, concessions, etc. At the same time, theoretically capital investments made by the owners of the subsoil are not inconsistent with the separation of the ownership of the subsoil and the ownership of land. The owner of the land comes to appropriate rent, while the capitalist investor appropriates normal profit.

Given the above property relations, the formation of social value involves a process of intra-industry competition. Depending upon the extent of differential productivity, there will be value transfers from one individual capital to another that would manifest themselves as differential rent. The internationalization of oil production is also an important part of this process. As we have pointed out in chapter three, during the early period of the oil industry in the Middle East, the production of oil was basically originated through the *formal* subsumption of labor under capital. However, as the material foundation of capitalism in the international oil industry developed, the production of oil has gradually assumed the characteristic of the *real* subsumption of labor under capital at the global level. In this manner, a social value at the level of international oil industry has been formed, and differential oil rent has emerged as a result of global competition.

Based on the analysis of the previous section, the same distinction should be made between the Ricardian or neo-Ricardian notion of the value of the marginal oil producer, and the notion of the *regulating market value* in the Marxian sense.[26] It is not always the case that the market value is the same as that of the marginal producer.

At the empirical level, we were able to identify the U.S. oil (lower 48 states) as the least productive region of the world.[27] The U.S. oil fields are also the most explored oil region of the globe. Since the effect of differential oil rent of the first type (DRI) cannot be separated from differential rent of the second type (DRII), determination of the least productive oil region in conjunction with the identification of *normal* size of capital investment, is an impossible statistical task. We have encountered the very same problem that Marx faced a little over a century ago in agriculture. Following Marx's method, we were able to identify the specific oil fields upon which the social value of oil was formed.[28] Due to the integration of oil production at the global level and the fact that the U.S. oil comes from the least productive oil fields, the social value associated with the above oil fields has become the social value of the international oil industry. Thus, it is through the decline in the productivity of the old U.S. oil fields located in the lower 48 states, that the magnitude of social value for the entire industry has to be increased. The sharp increase in the differential magnitude of this social value caused the entire oil industry to be subjected to an objective and fundamental reorganization and, therefore, *crisis.*

To sum up, the significance of Marx's theory of rent is in its specificity. Since rent is the effect of landed property in agriculture, Marx's theory of landed property figures significantly as its precondition. Likewise, an appropriate theory of oil rent should be related to the analysis of oil property relations.[29] Also the recent categorization of resources into renewable and exhaustible is a *superficial* distinction, and Marx's theory of rent in agriculture and the theory of oil rent developed in this volume are both theoretically and empirically far superior. For instance, unlike these categories which are associated with the concept of *pure competitive* equilibrium, (quite uncharacteristic of capitalism), Marx's theory of agricultural rent as well as the theory of oil rent are the result of the *actual* competition in capitalism.

NOTES

1. David Ricardo, *The Principles of Political Economy and Taxation* (London: J.M. Dent & Sons Ltd., Third Edition, 1976), p. 35, (emphasis added).
2. Alfred Marshall, *Principles of Economics* (London and New York: Macmillan and Co. Ltd., 8th Edition, 1961), p. 422.
3. Ricardo, p. 3.
4. Marshall, p. 144.
5. *Ibid.*
6. *Ibid.*, (emphasis added).
7. *Ibid.*, p. 147.
8. *Ibid.*, p. 156, (emphasis added).
9. Karl Marx, *Capital*, vol. 3 (New York: Vintage Books, 1981), pp. 751–52.
10. Marx, pp. 787–88.
11. *Ibid.*, p. 661, (emphasis added).
12. See the collected papers presented to the Symposium on Exhaustible Resources, held in London and published in *Review of Economic Studies*, 1974.
13. Ricardo, p. 38.
14. *Ibid.*, p. 56. However, Ricardo's chapter: 'On Machinery' is an exception to the above-mentioned way of thinking.
15. Karl Marx and Frederick Engels, *Selected Correspondence* (Moscow: Progress Publishers, 1975), pp. 47–48. This letter was brought to my attention by Anwar Shaikh.
16. Karl Marx, *Theories of Surplus Value*, vol. 2 (Moscow: Progress Publishers, 1968), pp. 42–43, (emphasis in original).
17. Ben Fine, 'On Marx's Theory of Agricultural Rent', *Economy and Society*, August, 1979, pp. 254–55.
18. *Ibid.*, p. 255.
19. Karl Marx, *Theories of Surplus-Value*, vol. 2 (Moscow: Progress Publishers, 1968), p. 273. See also *Capital*, Vintage edition, vol. 3, p. 798, Chapter 38.
20. Karl Marx, *Theories of Surplus-Value*, vol. 2, pp. 292–93, (emphasis in original).
21. It should be noted that absolute rent is not 'monopoly rent' in a sense that the neo-Ricardians, as well as those associated with the School of Monopoly Capitalism, have managed to interpret it. See Willi Semmler, 'Competition, Monopoly, and Differentials of Profit Rates: Theoretical Considerations and Empirical Evidence', *Review of Radical Political Economics*, Winter 1982, pp. 41–42.
22. Karl Marx, *Capital*, vol. 3, International Publishers' edition, p. 749, (emphasis added).
23. See the passage in which Ricardo talks about air, water, and the elasticity of steam. . ., on p. 39 of his *Principles*.
24. Anwar Shaikh, 'Differential Rent: Lecture No. 8', New School for Social Research, New York, p. 1. (Mimeographed).
25. For the analysis of the historical origins of oil rent in the Middle East, see Chapter III.
26. For instance see Bob Fitch, 'OPEC: The Big Cartel That Couldn't. . .', *Against the Current*, Spring 1982, pp. 16–22. The notion of oil rent used by this author is Ricardian although he produced no conclusive results.
27. See Chapters VII and VIII.
28. In order to identify the class of value-determining land in conjunction with regulating capital, Marx had to vary the price of production in agriculture. Confronted with a similar problem, we have to identify the characteristic of value-determining sphere of the oil production. If the increases of value and

prices are resulted from the decline in the productivity of the least productive oil fields, one has to be able to identify the types of investments that are made on these fields, and determine whether or not that would conform to this declining trend.

29. An example of the treatment of rent independent of the existing property relations in the oil industry can be clearly observed from Chevalier's 'oil rent'. In doing so, he designates five different oil rents, each of which relates to either a particular physical, or technological, or marketing aspect of oil production. These are: (1) 'quality rents', (2) 'position rents', (3) 'mining rents', (4) 'technological rents', and (5) 'monopoly rents', that in turn reflected the quality of crude oil, transportation cost, geological formation, technological differences, and the 'monopoly' of importing and exporting nations, respectively. The application of this approach is also visible from his uncritical acceptance of Marshall's distinction of *royalty* from *rent*, that is made on the basis of physical exhaustibility. See J.M. Chevalier, 'Theoretical Elements for an Introduction to Petroleum Economics', in *Market, Corporate Behaviour, and the State*, edited by A.P. Jacquemin and H.W. de Jong (The Hague: Martinus Nijhoff, 1976), pp. 288–289 and 297.

CHAPTER VI

THEORIES OF PURE COMPETITION AND COMPETITION IN CAPITALISM

1. Introduction

The oil crisis of 1973–74 was not only the crystalization of a fundamental change in the world oil industry, but it was and still is a testing ground that has shaken the foundation of orthodox and radical economics alike.

As was expected, following the crisis, the search for an 'adequate' explanation has led theorists to use the prevailing economic theories to their fullest extent. Nevertheless, given the demonstrated bankruptcy of neoclassical economics and the domination of *orthodoxy* in the remaining brands of economics as well, the cause of the oil crisis remained to be identified. To put it in Joan Robinson's words: 'The neoclassical heritage had great influence, not only on the teaching of economics but in forming public opinion generally, or at least in providing public opinion with its slogans. But when it comes to an actual issue, it has nothing concrete to say.'[1] The primary theoretical obstacle to developing a clear understanding of the crisis has been the dominance of the theory of monopoly-oligopoly borrowed from neoclassical economics. The theorists of the oil crisis have made the theory of oligopoly in general, and the notion of cartel in particular, the core of their analysis. That is why these 'old habits of thought' have prevented clear, scientific thinking. For 'The difficulty lies', Keynes once remarked, 'not in the new ideas, but in escaping from the old ones. . .'[2]

In the present chapter we will examine the orthodox economic theories of market structure in capitalism. It will be demonstrated that competition in capitalism is neither *pure* nor *imperfect*. In section 2, the theories of market structure associated with neoclassical economics will be presented. Section 3 will contain a brief exposition of Marx's theory of competition. The concept of pure competition, monopoly, and actual competition in capitalism will be examined in section 4. Finally section 5 will expose the striking resemblance of 'radical' and orthodox economics from the standpoint of Marx's theory of competition.

2. Orthodox Economics: The Economic Theories of Market Structure

There are four different types of markets in neoclassical economics: (1) purely competitive, (2) purely monopolistic, (3) imperfectly competitive, and (4) oligopolistic.[3] The first two models are about competition and monopoly in their *pure* forms, constituting an upper and the lower boundary of a *competitive spectrum,* where the last two models fall in between.

The model of pure or perfect competition is associated with the large number of tiny identical firms, entering and exiting the industry freely and instantaneously.[4] The firms of this category are engaged in the production *or* selling of standardized or homogenized products as they maximize their profits or minimize their losses through the equalization of their marginal cost (MC) and marginal revenue (MR). An additional feature of this model is the simultaneous equalization of MC and average revenue (AR). This will guarantee the condition of 'optimal resource allocation' or *efficiency.* Thus, it is said that production and exchange under pure competition would necessarily lead to both profit maximization and efficiency in the market economy.

In contrast, the model of pure monopoly refers to a one-firm industry. Therefore, conceptually, the question of entry and exit is irrelevant to this model. One rationale for monopoly rests on an assumption of *economies of scale.* It is argued that marginal and average costs decline indefinitely as the level of output expands.[5] Obviously, as the level of output expands, and the gap between the AR and MR curves widens, the output corresponding with profit maximization will substantially differ from the so-called efficient level of output. In other words, $MC = MR \neq AR$ is the consequence of imperfect competition in general, and pure monopoly in particular. The comparison of the above models leads to the conclusion that (1) competition is the *opposite* of monopoly, (2) the increase in the size of firms is a barrier to competition, and (3) the primary distinction of competition from monopoly is based on the number of firms in an industry.

With the formulation of the above conceptual framework, we can turn to the description of the last two models: imperfect (monopolistic) competition and oligopoly.[6] The model of imperfect competition is said to be relevant to an industry with a fairly large number of firms. Neither the size of the firms nor the types of products they make or sell are identical. There is a great deal of 'product differentiation' within the industry, but the size of the firms and their market shares are insignificant. Although the condition of entry and exit associated with this market model is not as free as the pure competition, entries and exits are easier than those of the oligopolistic markets.

In contrast with the purely competitive firms, an imperfectly competive firm faces a downward-sloping demand curve. Therefore, the equilibrium quantity of output associated with the condition of profit maximization is different from that of 'optimal resource allocation' or *efficiency.* As a result, no matter how small the size or how large the number of firms might be, imperfect competition, as opposed to pure competition, will always lead to an *inefficient* allocation of resources.

The remaining market model to be analyzed is oligopoly. The term oligopoly refers to a few sellers in the entire industry. Consequently, with the above conceptual framework in mind, one has to suspect that since there is a great deal of control over the prices and the quantities of goods produced by the individual firms, there is little competition in the industry. At the same time, there is no well-defined economic mechanism, or for that matter 'economic principle', that will govern the movements of prices and quantities of output in oligopolistic conditions. This situation stems from the nature of interactions among the firms in regard to the price and output decisions that are initially made by a firm or a group of firms in an industry. In other words, there is a great deal of *indeterminacy* in the outcome. The primary significance of this model is its nearness and virtual identity to the model of pure monopoly. In fact, the attribute of 'administrative pricing' is said to be evidence of monopoly power. With the existence of a few firms, price formation tends to become an *administrative* task, subject to the reaction of a few rivals. In this particular situation, the demand facing an oligopolistic industry becomes indeterminate. One cannot determine, or for that matter identify, a particular price in relation to a particular quantity for the entire industry without additional assumptions. Thus, the field of oligopoly is actually a special branch of Game Theory.[7]

3. Marx's Theory of Competition

In the previous section, an attempt was made to present the conceptual framework of orthodox economic theories of the market structure, i.e., perfect competition, imperfect competition, oligopoly and pure monopoly. The task of the present section is to present Marx's theory of competition as an alternative to the prevailing 'conventional wisdom' and show its significance in relation to his theory of value.

i. The Origin of Competition

In order to capture the essence of competition in Marx, one has to start from the level of capital as a whole and study the reproduction of the social relations of capitalism. In this connection, the notion of capital as a whole in relation to accumulation and the production of surplus-value is the starting point. As Marx put it, '*competition* is nothing other than the inner *nature of capital;* [here] its essential character, appearing in and realized as the reciprocal interaction of many capitals with one another, [that is] the inner tendency as external necessity'.[8]

The theoretical implications of the above passage in the context of Marx's theory of value can be summarized as follows:

1. The accumulation process under the rule of capital is a contradictory process.
2. Competition is the external manifestation of the above contra-

diction, and not the result of an arbitrary external force or motivation.
3. Competition does not originate in the sphere of circulation, i.e., from the level of many capitals. Thus, it cannot be eliminated through the elimination of competing capitals.
4. Competition is specific to the period of history under capitalism alone.
5. Competition, as the external manifestation of the motion of capital, is necessarily relevant to both *laissez faire* and *monopoly* phases of capitalism.
6. 'The coercive laws of competition'[9] lead to a merciless war of capital against capital for survival.

Recognition of the above points is essential in the examination of the intra-industry and inter-industry competition. Since competition among many capitals is merely *the effect* of 'the inner nature of capital', one has to start from the level of social capital-as-a-whole. Marx revealed that '. . . a scientific analysis of competition is possible only if we can grasp the inner nature of capital, just as the apparent motions of the heavenly bodies are intelligible only to someone who is acquainted with their real motions, which are not perceptible to the senses'.[10] Competition as recognized by Marx is thus a manifestation of capitalist production. Competition is the continuous process of *coercion* within social capital. Since competition is relevant to the sphere of production before its manifestation among many capitals, it is relevant to the entire period of capitalist production.

It should be noted that since competition is the 'inner nature' of capital, one cannot remove it from the analysis of capitalist society by way of assumptions.[11] If competition is the inner nature of capital, then with the further development of capitalism, the intensity of competition must be compounded. In this connection Marx noted that, 'As long as capital is weak, it still relies on the crutches of past modes of production, or of those which pass with its rise. As soon as it feels strong, it throws away the crutches, and moves in accordance with its own laws.'[12] Thus, there must be more competition in the era of the maturity of capitalism than its youthful age. 'Competition, since it derives from the inner nature of capital,' Weeks says, 'develops and intensifies as capital develops. When competitors are small and many, competition was primitive and embryonic. It is with the development of capital in its most advanced form, monopoly capital, that competition, too, develops to its fullest extent.'[13]

ii. The Mode of Manifestation of Competition

As we have pointed out earlier, the law of competition originates from the production process. Therefore, the pre-condition for what is called 'scientific analysis of competition' is the recognition of competition as the condition of accumulation. Having done that, we may now turn to the analysis of competition among many capitals, both within and between industries.

The commodities produced by individual producers within an industry are often subject to different techniques and costs of production. The existence of such differences, therefore, leads to the formation of different *individual values* associated with different individual units of production in an industry. The individual producers with higher technique and more efficient conditions of production are the ones who have smaller individual values as opposed to those with lower productivity. Marx maintains that:

> The law of the determination of value by labor-time makes itself felt to the individual capitalist who applies the new method of production by compelling him to sell his goods under their social value; this same law, acting as a coercive law of competition, forces his competitors to adopt the new method.[14]

However, as long as the more efficient techniques of production are not yet generalized, given the formation of social value on the average technique, the individual capitals associated with these higher techniques will make higher rates of profit through competition. Thus, intra-industry competition will always lead to differential rates of profit—associated with differential productivities of individual capitals—within an industry. At the same time, it is through intra-industry competition that all the individual units of capital will stand in reciprocal relation to each other, and form a social value for the entire industry. It should be pointed out that the above social value is based on the average technique of production in an industry.

In contrast, competition at the inter-industry level will lead to the equalization of profits with the transformation of values into prices of production. Marx explicitly asserts that:

> What competition brings about, first of all in one sphere, is the establishment of a uniform market value and market price out of the various individual values of commodities. But it is only the competition of capitals in *different* spheres that brings forth the production price that equalizes the rate of profit between those spheres.[15]

It should be emphasized that the existence of differential profitability in an industry is not due to competition. But competition is the condition of the transformation of the existing differential productivities of individual capitals into differential profitability in a single industry. In this connection Anwar Shaikh correctly observes, that 'Whatever the reason for the differential productivity, it will manifest itself as differential profitability.'[16]

As we have pointed out, inter-industry competition will lead to equalization of the rates of profit. However, there should be a distinction between the above equalization which is *tendential* and the notion of *equilibrium* in orthodox economics. The equalization of the rates of profit involves the movement of capital from one industry to another. Also since 'in every sphere of actual production. . . the transfer of capital from one sector to another presents significant difficulties', it induces rising profit rates in some industries and falling profit rates in others, at any particular time.[17] Nevertheless, the above fluctuation will be evened out over a 'cycle of lean and fat years'.[18] Thus, in contrast with orthodox economics, Marx's theory of competition does not lead to an *equilibrium* of the rates of profit.[19]

The consequence of the formation of differential rates of profit in a single industry, and the tendency to equalization of the rates of profit between industries through competition is the transfer of value both within and between industries. The amount of this value transfer depends upon the differential magnitude of individual values and the social value within a single industry and the difference between the prices of production and market values in the entire economy.[20] Thus, there is a possibility for the transfer of value within a transnational industry such as the international oil industry. In this case such a transfer is based on the differential productivities of individual oil regions and their transformation to differential profit rates through competition.[21] However, the actual process of value transfers primarily depends upon the degree of the internationalization of capital in one or more industries. Moreover, the expansion of capital and the impact of what Marx calls the *propagation* of capitalist mode of production *'through international competition'*, is historically prior to the existence of full-fledged intra-industry as well as inter-industry value transfers at the international level.[22] As it turns out, the above process leads to an intense struggle among capitals. This struggle often turns into a war for survival rather than a war for the appropriation of the average rate of profit. To paraphrase Shaikh, 'this average rate is, after all, an average of the outcomes of hundreds of thousands of battles fought over varying terrain and with varying weapons—you pay your money and you take your chances.'[23]

To sum up: we have pointed out that the phenomenon of competition in Marx is primarily the reflection of the inner nature of capital. Since the social capital-in-general can only reproduce itself through the multitude of independent fractions of capitals, the mode of manifestation of competition is the competition of many capitals. Competition of many capitals can take place at two different levels: (1) through intra-industry competition, and (2) through inter-industry competition. The first form of competition leads to differential rates of profit within a single industry, while the second form of competition results in the *tendential* equalization of the rates of profit. Both cases involve the transfers of value from one individual capital to another. This implies that competition is the war of capital against capital for survival.

4. Perfect Competition and Competition in Capitalism

As we have seen, perfect or pure competition in orthodox economic theory refers to an ideal situation where there are infinite numbers of firms of infinitesimal size *coexisting* with each other. These firms are all 'price-takers' and faced with perfectly elastic demand curves. As a result, they do not exert any influence on the market price or on each other. In addition, the condition of entry and exit is absolutely free, and the firms can enter and leave the industry free of cost, instantaneously. Actually, 'perfect competition can exist only when product, at every level of output, is carried on under constant returns to scale, [with] no fixed cost and, hence, no optimum size of the plant'.[24]

Perfect competition is really a situation in which a great number of microscopic agents called firms coexist passively with one another. Conceptually such a competition is nothing but the lack of competition and abundance of harmony. Empirically, such a situation pretends to be a bastion of cooperation. Yet there are no relationships among the firms; they are mysteriously isolated from each other.

According to the fantasy of perfect competition, any minor influence of one firm upon another is characterized as imperfect competition. Thus, from the viewpoint of neoclassical economics, the ideal capital is always 'non-capital'.[25] The theory of perfect competition is the mystification of the process of concentration and essentially centralization of capital through the substitution of harmony for contradiction, isolation for socialization, fancy for reality, and, finally, submission for competition. The reality of competition in capitalism shows that, in fact, the competing capitals do influence each other; they do influence the scale of production; they do actually engage in driving each other out of business; and finally, they do often take over each other's business. It is indeed the very essence of competition that leads to concentration and centralization of capital and in turn to a further intensification of the competitive struggle among capitals.

As we pointed out above, pure monopoly is another mental construct developed by neoclassical economic theory, which is not something separate from pure competition. In fact, the theory of pure monopoly is the *logical extension* of pure competition. From the viewpoint of *real* competition in capitalism, while the impossibility of competition in pure competition is due to the passive character of the firms, the impossibility in pure monopoly is guaranteed by the identity of the firm and industry.

It should be noted that competition in orthodox economics always originates from the sphere of exchange. Therefore, the effect of actual competition (i.e., centralization and concentration of capital) is seen as the removal of competition. If competition is *pure* and the competing firms are operating on the basis of constant returns to scale with no fixed cost, it is not surprising that the phenomena of concentration and centralization of capital are seen as 'monopoly'.

Another problem associated with orthodox economic theory is the distinction between imperfect competition and oligopoly. It is argued that imperfect competition is a special case of oligopoly if the interactions among firms are recognized.[26] If so, depending upon the courses of action taken by one or more firms, and the subsequent reaction of others, output and prices will be varied. In fact, according to orthodox economics there is no determinate solution to such problems. Thus, one has to generalize the case of oligopoly for the entire spectrum of imperfect competition. It turns out that the only applicable theory is Game Theory. Nevertheless, it is interesting to note that Game Theory deals with the analysis of power and warfare, and it is no accident that the theorists such as J.V. Neumann and O. Morgenstern approached economics in these terms as early as 1944.[27]

5. *Radical Economics and the Prevailing Orthodoxy*

In the previous sections we critically examined a number of theoretical implications of the *ideology* of pure competition in conjunction with other types of market structures, such as oligopoly. We pointed out that the lack of distinction between the *essence* and the *phenomenal forms* has led orthodox economics to distort the meaning of competition in capitalism. As for the types of market structure associated with neoclassical economics, it is said that the model of pure competition is the basis of all.

The influence of the above way of thinking is also evident from neo-Ricardian economics which 'associates perfect competition with "competitive capitalism" and imperfect competition with "oligopolistic capitalism" '.[28] However, one may wonder at the growing distance of 'Marxian' economics from Marx's economics. The following passage from *Monopoly Capital* may be worth a thousand words in showing what we mean by 'Marxian' economics:

> We must recognize that competition, which was the predominate form of market relations in nineteenth-century Britain, has ceased to occupy that position, not only in Britain but everywhere else in the capitalist world. Today the typical economic unit in the capitalist world is not the small firm producing a negligible fraction of a homogeneous output for an anonymous market but a large-scale enterprise producing a significant share of the output of an industry, or even several industries, and able to control its prices, the volume of its production, and the types and amounts of its investments.[29]

Here a small firm of negligible production, i.e., a 'price-taker' in a purely competitive market is said to be the basis of 'competitive capitalism', whereas the large-scale enterprises with a substantial amount of market shares and considerable 'monopoly power' characterize the era of 'monopoly capitalism'. A simple inspection of the above passage will reveal the extent to which the orthodox economic thinking has penetrated into our 'Radical' economics of today. It is true that the emerging objective realities of our time have led to full-scale bankruptcy of orthodox economics in general, and neoclassical economics in particular, but unfortunately the orthodoxy is a living legend that persistently survives through a brand of economics that claims to represent its very opposite.

A very important distinction between Marx's economics and 'Marxian' economics is the analysis of value and within its core, theory of competition. For Marx, competition is the condition of accumulation. As a result, competition is relevant to both *laissez-faire* and *monopoly* capitalism. Moreover, in Marx, competition is not *pure* or fictional but the process of war for the survival of capital. For Marx, the big sizes and small numbers of firms in an industry are not an indication of the non-existence of competition.

6. Conclusion

We have argued that the theory of pure competition is an ideal construct which actually projects the condition of subordination rather than competition. It should be apparent by now that competition is the essential characteristic of the capitalist mode of production. Thus, it is relevant to the entire period of capitalism. In addition, the concentration and centralization of capital lead to further competition, because competition is the war of capital against capital.

Competition at the level of many capitals will lead to differential rates of profit within individual industries and equalization of the rates of profit between industries. Both of the above processes involve the transfer of value from one individual capital to the other. The basis of the transfer of value through intra-industry competition is the existence of differential productivity which leads to the condition of differential profitability. The transfer of value can also take place within an international industry through competition. Competition in the international oil industry was an illustrative case during the world oil crisis of 1973-74. Here, the formation of newly-established value, i.e., regulating market value, on the one hand, resulted in differential productivity in different oil regions of the world, and on the other, led to significant value transfers in that industry. However, due to the existence and intervention of oil property relations in this particular industry, the resultant differential profitabilities were transformed into differential oil rents. Thus, contrary to the prevailing opinion, the formation of oil rents emerged through competition rather than through monopoly.

NOTES

1. Joan Robinson, *Economic Philosophy* (Garden City, New York: Anchor Books, 1964). p. 131.
2. John M. Keynes, *The General Theory of Employment, Interest and Money* (New York: Harcourt, Brace & World Inc., 1964), p. viii.
3. See any standard economics textbook such as Samuelson's *Economics*, 9th Edition, Chapters 25 and 26. For theoretical sources, see E.H. Chamberlin, *The Theory of Monopolistic Competition* (Cambridge, Mass.: Harvard University Press, 1950); Joan Robinson, *The Economics of Imperfect Competition* (London: Macmillan, 1933); Piero Sraffa, 'The Laws of Returns under Competitive Conditions', *Economic Journal* vol. 36 (1926), pp. 535–550; and George J. Stigler, 'Perfect Competition, Historically Contemplated', *Journal of Political Economy*, vol. 65 (1957), pp. 1–17.
4. Throughout this chapter, 'pure' and 'perfect' are used interchangeably.
5. Within the framework of the neoclassical notion of competition and its corresponding categories of market structure, some theorists attempt to explain the oil crisis through *marginal* analysis. See J.M. Chevalier, 'Theoretical Elements for an Introduction to Petroleum Economics', in *Market, Corporate Behaviour, and the State*, edited by A.P. Jacquemin and H.W. de Jong (The Hague: Martinus Nijhoff, 1976), pp. 290–293.
6. The notions of monopolistic and imperfect competition are used interchangeably throughout this chapter.
7. Donald Dewey, *The Theory of Imperfect Competition* (New York: Columbia University Press, 1969), pp. 16–20, 158–64, and 192–93.
8. Karl Marx, *Grundrisse* (New York: Vintage Books, 1973), p. 414, (emphasis in original).
9. Karl Marx, *Capital*, vol. 1 (New York: Vintage Books, 1977), p. 436.
10. *Ibid.*, p. 433.
11. John Weeks, *Capital and Exploitation* (Princeton, New Jersey: Princeton University Press, 1981), p. 153.
12. Karl Marx, *Grundrisse* (New York: Vintage Books, 1973), p. 651.
13. John Weeks, *Ibid.*, p. 167. See also James Clifton, 'Competition and the Evolution of the Capitalist Mode of Production', *Cambridge Journal of Economics*, vol. 1, no. 1, 1977, pp. 137–51.
14. Karl Marx, *Capital*, vol. 1, Vintage edition, p. 436.
15. Karl Marx, *Capital*, vol. 3, Vintage edition, p. 281, (emphasis in original).
16. Anwar Shaikh, 'Lecture No. 8: Differential Rent', Department of Economics, New School for Social Research, New York, p. 1. (Mimeographed.)
17. Karl Marx, vol. 3, Vintage edition, pp. 310–11.
18. *Ibid.*, p. 311. Also Anwar Shaikh, 'Lecture Notes No. 9', where he extends Marx's tendential equalization of the rates of profit to include the condition of entry, condition of exit, and dynamics of the transfer of values between industries.
19. Also see John Weeks, *Capital and Expolitation* (Princeton, New Jersey: Princeton University Press, 1981), p. 172. His chapter on competition includes an interesting discussion of the law of uneven development.
20. Anwar Shaikh, 'Lecture Notes No. 8: Differential Rent', p. 4 (Mimeographed); this approach, i.e., the formulation of conditions of competition both for industry and agriculture in respect to intra-industry and inter-industry value transfer, is

quite original in Shaikh's analysis. He also, following Marx, extended the theory of competition to the condition of entry and exit of capitals. See also Willi Semmler, 'Competition, Monopoly, and Differential Profit Rates—A Reconsideration of the Classical and Marxian Theories', *Rivista Internazionale di Scienze Economiche e Commerciali*, Agosto, 1982.

21. This process involves the relation of oil property and oil rent as its phenomenal form which results in the *permanent* transformation of differential profitability to differential oil rent.
22. Karl Marx, *Grundrisse*, Vintage edition, pp. 729–30 (emphasis added).
23. Anwar Shaikh, 'Neo-Ricardian Economics: A Wealth of Algebra, A Poverty of Theory', *Review of Radical Political Economics*, vol. 14, Summer, 1982, p. 77.
24. Donald Dewey, *The Theory of Imperfect Competition: A Radical Reconstruction* (New York: Columbia University Press, 1969), p. 192.
25. Anwar Shaikh, 'Lecture Notes No. 6', Department of Economics, New School for Social Research, New York, 1981, p. 9. (Mimeographed.)
26. Donald Dewey, pp. 192–93.
27. John Von Neumann, and Oskar Morgenstern, *Theory of Games and Economic Behavior* (Princeton, New Jersey: Princeton University Press, 1944).
28. Anwar Shaikh, 'Neo-Ricardian Economics: A Wealth of Algebra, A Poverty of Theory', *Review of Radical Political Economics*, vol. 14, Summer, 1982, p. 80.
29. P. Baran and P. Sweezy, *Monopoly Capital* (New York: Monthly Review Press, 1966), p. 6. This particular critique should not, by any means, create an impression that the intention here is to downplay Sweezys' lifelong contributions to political economy, as one of the leading theorists of this century. It goes without saying that the younger generation of radical economists and progressive political activists of today is greatly influenced by his teachings.

CHAPTER VII

AN ANALYSIS OF THE UNITED STATES OIL PRODUCTION

1. Introduction

It has been hypothesized that the oil crisis of 1973-74 was the consequence of an increase in the long-term production costs of oil in the United States prior to October, 1973. It has also been indicated that due to the nature of the industry, and the existing social relations of production associated with it, historically the U.S. oil region has become the least productive area of the world. In addition, we maintained that as a result of the integration of oil production at the global level, values, oil rents and market prices tend to be formed at the level of the international oil industry. Table 10 shows the investment costs necessary to produce a new daily barrel of oil in various oil producing nations. It is clearly evident that the production of oil in the U.S. is by far the most expensive in the world. Thus, it is reasonable to proceed by focusing our research on the United States oil industry. When the least productive producer becomes even less productive (depending upon the conditions under which generalization of least productive conditions takes place), it leads to the reorganization of production by way of crisis, which would result in the formation of new value, a new rent structure, and new prices of higher magnitude.

Table 10

*Approximate Cost of Capital Investments of New Daily Barrels of Crude Oil**

Country	*1960-1972*
Canada	3,100
Indonesia	1,400
Libya	200
Nigeria	800
Saudi Arabia	200
United Kingdom	3,500
United States	3,700
Venezuela	1,000

* All figures in U.S. dollars.

Source: Frank R. Wyant, *The United States, OPEC, and Multinational Oil* (Lexington, Mass.: D.C. Health and Company, 1979), p. 117, Table 5-13.

We shall demonstrate that the theoretical arguments advanced so far, along with the empirical expositions presented in the present chapter, will corroborate the hypothesis and support our theory of the oil crisis. Our task here is to show that the retention of the aged oil fields in the United States, and the resultant decline in the average oil recovery, was the underlying factor in reorganization of production, and formation of new value, market price, and rent (royalties) in the international petroleum industry. This reorganization ultimately led to the crisis of 1973-74 which, as we have seen, did not confine itself to one country or two but swept through the whole global structure.

2. *Evaluation of the Statistical Information*

Although a great deal of information exists on geology and the exploration and production of crude oil in the United States, there is not enough information available on investment and productivity. This may be one reason, among others, that fundamental research on the present subject is so limited. Nevertheless, there are rare instances when one may have the good fortune to come across information which is useful, meaningful and adequate in connection with the oil crisis of 1973-74.[1]

Although the figures for the daily average of oil production in the entire United States and for the individual producing states have been published for some forty years, the extent of output of the oil production in connection with each depth category of the principal oil-producing states had not been properly disaggregated and identified. It is only after the publication of the three technical papers cited above that one actually has enough statistical information on particular years. Since the aforementioned sources deal with the evaluation of conditions of oil production, in 1965, 1971, and 1974, only the period relating to these years can be empirically used for our analysis. In other words, although the time series data are not available, it is possible to compare the conditions associated with the 1965-71 and 1971-74 periods.

3. *Recognition of Different Spheres of Oil Production*

There are three different ways by which additional oil reserves can possibly be developed and made available for production in the oil industry. These are: (1) uncovering new discoveries, (2) extending old discoveries, and (3) recovering additional oil from the existing oil fields. It goes without saying that the above spheres should be distinguished from one another because the conditions of capital investment associated with each one are significantly different in terms of capital intensity. While useful in principle, the statistics underlying producing rates, along with the classification of domestic oil reservoirs provided by the above sources, cannot provide a clear-cut formulation which separates the spheres connected with the extension of old oil discovery and enhanced oil recovery. But, fortunately, we need not be too concerned about this. There is a general consensus that the deeper oil wells, which at the same time produce more than 32 barrels a day, are considered to be the likely candidates for *extension* as compared with the more shallow ones of lower productivity. The latter category usually is the subject of the application of enhanced oil recovery methods. In addition, there is the sphere of new oil discovery which is associated with a higher degree of productivity during the crisis period. The significance of the above classification will become clear when we analyze the long-term investments per barrel of oil production in the United States during the period of crisis.

4. *Intensification of Production and the Reversal of the Oil Recovery Trend*

More than 90 per cent of the U.S. crude oil was produced from reservoirs located in nine principal states during the period of 1965–74. From this, nearly two-thirds were produced from Texas and Louisiana combined. The remaining one-third came from California, Oklahoma, Wyoming, New Mexico, Alaska, and Mississippi.[2]

The comparison of trends of oil recovery for the entire United States and the principal oil producing states, on the one hand, and that of the long-run investment per barrel, on the other, is one possible way to examine the actual arrangement of the emerging conditions that led to the formation of a new cost structure, through crisis, in the oil industry. The identification of decline either in the trend of oil recovery in the old oil fields, or in the rate of new oil discoveries, in conjunction with the long-term investment trends associated with these spheres of production during the 1971–74 period, reveals which area has a dominance in regulating the restructured industry's market value and market prices.

A comparison of these trends requires calculation of the average oil production per well for the above-mentioned major oil producing states of the United States for 1965, 1971, and 1974. But the average of these averages does not represent the true meaning of average. Appropriate weights, i.e., the actual shares of production in these states, must be employed if the average is to have economic meaning. Table 11 shows the average oil recovery and the shares of oil production for major oil producing states of the U.S. for 1965, 1971, and 1974.

Since the scope of our analysis includes the old oil reservoirs located in the lower 48 states of the United States and since Alaskan oil is among the most productive regions in the United States there seems to be no need for inclusion of the state of Alaska in the present analysis. At the same time, Illinois, a major oil producing state in 1965, lost its position in 1971 as well as in 1974. Therefore, for the sake of uniformity, it seems advisable to disregard the amount of production related to Illinois. What is left for comparison is the weighted average of oil recovery for the eight remaining

Table 11

Daily Average Oil Recovery per Well and Production Shares of Major Oil Producing States for 1965, 1971, and 1974

Major Oil Producing States	*1965* Average Oil Recovery***	*1965* % of Production Share	*1971* Average Oil Recovery	*1971* % of Production Share	*1974* Average Oil Recovery	*1974* % of Production Share
Texas	12.82	36.0	18.1	36.5	21.0	40.7
Louisiana	46.88	20.0	69.7	25.8	62.3	21.3
California	21.13	12.2	24.3	11.0	21.8	10.7
Oklahoma	7.37	8.1	8.0	5.8	6.4	5.6
Wyoming	52.18	5.2	56.9	4.6	47.5	4.6
New Mexico	21.92	4.4	22.3	3.4	19.4	3.1
Alaska*	–	–	1240.5	2.5	951.6	2.4
Kansas	6.28	4.0	6.8	2.3	4.0	2.0
Mississippi	51.69	2.1	52.6	2.0	48.1	1.7
Illinois**	6.03	2.5	4.1	1.2	3.2	0.9
		94.5%		95.1%		93.0%

* Alaska was not considered as an oil producing state in 1965.
** Illinois is not among the nine principal oil states in 1971 and 1974.
*** Average daily oil recovery per well in barrels.

Source: U.S. Department of the Interior, 'Depth and Producing Rate Classification of Oil Reservoirs in the 14 Principal Oil Producing States', Bureau of Mines Information Circulars 8362 (1967), and 8675 (1975A); and U.S. Department of Energy, *Depth and Producing Rate Classification of Domestic Oil Reservoirs,* 1974 (Washington, D.C.: Government Printing Office, 1978).

states for 1965, 1971, and 1974.

In Table 11, the first column relates to the weighted average oil recovery per well for the above years. These weighted averages are associated with the reservoirs located in the eight major oil producing states of the U.S. Simple inspection of this trend reveals that the oil recovery per well substantially increased during 1965-71 and declined in the subsequent period. During the period of 1971-74, average oil recovery in Texas, in spite of the individual trends of other major producing states, increased. Yet such an increase did not reverse the general downward tendency.

Tables 12 and 13 formulate summaries of the average oil recovery trends under different circumstances. If we eliminate the oil production in Texas from our analysis, the percentage changes in average oil recovery per well for the periods of 1965-71 and 1971-74 show a slight difference. As Table 13 shows, during the 1965-71 period the average oil recovery for major oil producing states, including Texas, increased by 46.6 per cent, while the same average with the exclusion of the state of Texas rose about 49.2 per cent. For the period of 1971-74, the corresponding figure for the average of all major oil producing states declined by 11.0 per cent, while the average oil recovery for all but Texas shows a decline of 14.4 per cent. The point here is that there has been a consistent trend in the average oil recovery, regardless of inclusion or exclusion of Texas, which exhibited an increase in oil recovery per well for the period of 1971-74. This point is significant because nearly two-fifths of the U.S. oil production comes from the Texas oil fields.

Table 12

The Weighted Average of Daily Oil Recovery per Well in Major Oil Producing States

Year	*Weighted Average Oil Recovery*	*Weighted Average Oil Recovery for All With the Exclusion of Texas*	*Average Oil Recovery Adjusted for MDF**	*Average Oil Recovery Adjusted for MDF With the Exclusion of Texas*
1965	24.11	31.37	32.91	42.92
1971	35.35	46.82	38.21	50.94
1974	31.43	40.09	31.43	40.09

* MDF stands for Market Demand Factor.

Source: Compiled by the author from U.S. Bureau of Mines, Information Circular 8362 and 8675; U.S. Department of Energy, DOE/EIA-0097, October 1978.

Table 13

Average Daily Oil Recovery Trend of the U.S. Oil, 1965-1974

	1965-71	*1971-74*
Trend of Average Oil Recovery Per Well	+46.6 %	- 11.0 %
Trend of Average Oil Recovery Per Well Adjusted for MDF	+16.1 %	- 17.7 %
Trend of Average Oil Recovery Per Well for all Major Producing States except Texas	+49.2 %	- 14.4 %
Trend of Average Oil Recovery Per Well Productivity Adjustment for MDF (except Texas)	+18.7 %	- 21.3 %

Source: Compiled by the author from *Bureau of Mines, Information Circular 8362 and 8675*; and U.S. Department of Energy, DOE/EIA-0097.

Another important consideration in the calculation of the average oil recovery is that the oil production of some major producing states was regulated under the mechanism of *Market Demand Factor.* Market Demand Factor (MDF) is the ceiling on production over and above the forecasted demand. Historically this production restriction was partly a response to the huge increase in the amount of oil from a newly discovered East Texas field by the major oil companies, in order to control the amount of oil production and stabilize the oil prices. Fragmentation of the oil leases is another dimension of this problem. Thus, MDF represents a fraction of the volume of potential oil production that is allowed to be marketed.[3] This mechanism usually fixes the amount of production as a certain percentage of the potential capacity of oil reservoirs in some states. Thus, the *potential* recovery of such oil wells differs from their *actual* oil recovery. In this connection, Texas and Louisiana are the ones in which MDF has been substantially in effect. The figure for 1971 for the above states, i.e., Texas and Louisiana, was on the average about 73 per cent of potential production. (Production based on the MDF was actually discontinued by 1972.)

It is apparent that the impact of the above regulation in the calculation of average oil recovery per well for 1965 and 1971 should be discounted. In other words, the figures of potential oil recovery have to be substituted for recovery ratio of the reservoirs which were regulated by the MDF. For this purpose, knowing the percentage of total production of regulated oil in the producing states in which such a regulation is predominant, is important.

In order to tackle the above problem, we have to examine Texas and Louisiana. In these states, the regulation of production was substantial and the application of MDF was widespread. With reference to the decision on the portion of production which has been under the MDF, we only had access to data on the division of reservoirs into primary and artificially stimulated production that were published in 1965. Since the MDF is usually used to regulate the more productive reservoirs in order to prolong their life span, or to prevent their early depletions, which often results from overproduction, it seems plausible that it is generally applied to the primary producing reservoirs as opposed to the stimulated ones.

By making use of the percentage of oil from primary producing reservoirs in 1965 and 1971 subject to MDF calculation, we were able to come up with a far more reasonable figure of average oil recovery per well for 1965 and 1971. Table 12 exhibits the daily average of the U.S. oil recovery per well adjusted for MDF for the related years. It also includes the figures (exclusive to Texas production) with the consideration of MDF for Louisiana. The objective of the above adjustment is to find out what the rate of potential oil production, i.e., production in the absence of such restrictions, has been for the years of 1965 and 1971.

With the above findings, it is clearly evident that the conditions of capital investment and production were entirely different during the 1971-74 period as compared to that of 1965-71. A simple inspection of Table 13 reveals that under all the circumstances explained above, the period of 1965-71 shows a substantial increase in oil recovery per well, while the corresponding results for the period of 1971-74 are consistently negative. Moreover, during 1971-74, as opposed to the previous period, the increasing magnitude of capital investments on the existing oil fields, and the production of reserves by way of intensification, extension, and enhanced recovery methods, ultimately led to the subsequent decline of average oil recovery per well in the old U.S. oil fields.[4]

One has to bear in mind that such a decline in the average oil recovery per well in the U.S. oil fields is not incidental to the oil crisis of 1973-74. The crisis is, in fact, both the symptom and the social mechanism for the generalization of production under the newly emerged conditions. During this period the magnitudes of capital investments per barrel and the individual value formed on the U.S. aged oil fields, due to the intensification of production, had significantly increased. At the same time, since the above oil fields constitute the least productive oil regions of the United States, the newly formed individual value on these fields tends to become the basis of social value through competition in the entire industry.

5. *Capital Investments and Individual Values in Different Spheres of Oil Production*

In the present section it will be shown that the total capital expenditures per barrel of oil in the United States tripled during the period of 1971-74. During the same period, the U.S. oil price also tripled. In this particular analysis, the 1971-74 period provides the framework for understanding the 1973-74 oil crisis.[5]

The establishment of the relationship between the U.S. development expenditure per barrel and the U.S. wellhead price of oil is a basis for the argument that the increase in the price of oil is greatly influenced by an increase in the U.S. capital costs. Given the fact that the U.S. oil region is the least productive oil region in the world, one might expect that the oil price, at the level of the international oil industry, will be regulated by the U.S. cost structure. In other words given the high degree of internationalization, the U.S. oil value becomes the *regulating* market value of oil for the entire international oil market.

However, it is interesting to be able to isolate the source of the above cost increase. This can be done by separating exploration from development capital costs, i.e., the two components of total investments. Having done that, it will be demonstrated that, in fact, the primary source of the cost increase is U.S. development costs. In other words, the increase in the long-run cost of oil development from the older U.S. oil fields, rather than the long run investments of new discoveries, was the principal source of the declining U.S. oil productivity.

Table 14 shows the trend of total U.S. capital expenditures per barrel, exploration expenditures per barrel, and development expenditures per barrel for the period of 1966–75. The total capital costs per barrel include the exploration and development costs per barrel. The exploration costs refer to the capital costs assocaited with the oil recovery from the newly discovered oil fields or the new reservoirs within the old oil fields. Meanwhile, the development costs include the capital investments made to recover oil from the aged oil fields. Such an oil recovery is called reserve addition through *'extensions'* and *'revisions'* in the oil literature.

Table 14

*Oil Price and Different Categories of the Capital Expenditures Per Barrel in the U.S. Oil Industry**
1966–1975

Year	*Wellhead Price of Oil*	*Total Capital Costs Per Barrel*	*Development Capital Costs Per Barrel*	*Exploration Capital Costs Per Barrel*
1966	2.88	1.138	0.984	2.456
1967	2.92	1.190	1.048	2.270
1968	2.94	1.770	1.579	2.886
1969	3.09	1.929	1.646	4.072
1970	3.18	1.193	1.005	2.209
1971	3.39	1.272	1.075	3.987
1972	3.39	2.897	2.787	3.404
1973	3.89	2.308	2.068	4.590
1974	10.13	3.953	3.885	4.321
1975	12.03	5.319	5.769	3.797

* All the figures are in U.S. dollars.

Source: API, *Reserves of Crude Oil, Natural Gas Liquids, and Natural Gas in the United States and Canada as of December 31, 1977,* vol. 32, June 1978; API, *Petroleum Basic Data Book,* various years.

In Table 15, the highest amount of increase is associated with the column showing the development capital costs per barrel during the period of 1971-74. This increase, i.e., the increase in the development costs per barrel, rose from $1.075 to $3.885, or more than tripled. However, the increase in the trend of exploration costs per barrel is only 8 per cent, an insignificant increase by comparison. The total capital costs per barrel also more than tripled during the 1971-74 period. The changes that took place in the trend of all the categories of costs during the period of 1966–70 are nevertheless by far smaller. (See Table 15.)

Table 15

The Changes in the Trends of the Oil Prices and U.S. Capital Expenditures Per Barrel 1966–1975

	Percentage Change During 1966–70	*Percentage Change During 1971–74*	*Average Per Barrel 1966–71*	*Average Per Barrel 1972–75*	*Change in the Average of the Two Periods (1966–71; 1972–75)*
Exploration Costs Per Barrel	10% (decline)	8.4% (increase)	$2.98	$4.03	35% (increase)
Development Costs Per Barrel	7% (increase)	261%(increase)	1.23	3.63	195% (increase)
Total Capital Costs Per Barrel	5% (increase)	211%(increase)	1.41	3.62	156% (increase)
Wellhead Oil Price	10.4% (increase)	199%(increase)	3.07	7.36	140% (increase)

Source: Compiled from Table 14 by the author.

The comparison of the changes that have taken place during the 1971-74 period with the changes in the average costs associated with the periods of 1966-71 and 1972-75, demonstrates that there is a significant increase in the development costs and total costs per barrel. In fact both the development costs per barrel and total costs per barrel more than tripled during the period of 1971-74. This result is in conformity with the tripled level of oil prices during the same period. It should be emphasized that in the main the measure of productivity used throughout this study is investment per barrel. This measure is referred to as capital-output ratio in the economics literature. The availability of statistical data for various oil regions of the world among other factors has made the usage of this measure widespread in the oil industry.

In this connection, it is evident that (1) the more than threefold increase in development costs per barrel partly coincided with the period of the oil price increase, and (2) the increase in the exploration cost per barrel is not in conformity by comparison. Moreover, this increase, i.e., an 8.4 per cent increase in the average exploration costs during the period of 1971-74, corresponds with a 15 per cent increase in oil discovery. (See Table 16.)

Table 16

Total New Field Exploration and New Oil Discovery Rates for the United States (Excluding Prudhoe Bay, Alaska) (1965-1975)

Year	*New Oil Discovery Per Foot* (barrels of oil)*			*TOTAL New Field Exploration Per Foot** (barrels of oil equivalent*
1965	18.64			–
1966	14.33	1966-71		23.00
1967	16.49	average:		26.01
1968	14.09	13.55		14.20
1969	9.71		percentage	12.14
1970	18.39		change:	19.52
1971	8.28		15%	13.08
1972	14.26	1972-75		13.11
1973	14.22	average:		16.89
1974	17.68	15.58		18.15
1975	16.18			18.70

* It is referring to crude oil only.
** It is referring to oil and gas together.

Source: Compiled by the author from API *Basic Petroleum Data Book* and U.S Department of Energy, *An Analysis of the Productivity of Domesti Petroleum Exploration Activities* (Washington, D.C.: Government Printin Office, 1978).

The development costs are the capital expenditure made on the aged U.S. oil fields. Consequently, the correspondence of the fourfold increase in the 'posted price', and the threefold increase in the U.S. oil price in conjunction with the more than threefold increase in development costs can explain the significance of *aged oil fields* in the crisis and the reorganization of production thereafter. It can be inferred that the increase in the cost structure of these oil fields led to the establishment of a new value of a larger magnitude for the U.S. oil industry. Since the U.S. region is the least productive oil region of the world, the U.S. oil value became the value norm for the international oil industry. As we pointed out earlier, the above process has evolved through competition among different individual values produced in different oil regions of the world.

We have demonstrated that the average oil recovery of oil fields in the U.S. regardless of prevailing conditions such as the existence of a regulating mechanism of Market Demand Factor (MDF) in some producing states, increased substantially during the 1965–71 period, and declined during the period of 1971–74. It is probable that the concentration of capital investment on the previously discovered oil fields, and the production of reserves by way of the *extension* of old oil fields, led to the decline of average oil recovery in the United States. Thus, one has to perceive that there were two distinct possibilities for the reorganization of production within the entire industry: (1) let the majority of old oil fields in the U.S. be abandoned and excluded from production in order to preserve the *old value and the price structure*, or (2) let the prevailing conditions in such oil fields, i.e., conditions associated with the decline of average oil recovery, be generalized for the entire industry. In this case, we have already observed that a new price structure has emerged. It is our contention that the second *possibility* became an *actuality* during the 1973–74 crisis.

6. The Role of New Oil Discoveries

The production of oil reserves can actually take three different forms. We have dealt with two of these possible forms, i.e., extensions and revisions, in previous sections. The status of the remaining form of reserve production, or new oil discovery, can be dealt with briefly in the present section. First of all, we need to follow the same pattern of periodization as we have in the previous section, for the sake of uniformity and and possibility of comparison. Thus, the trend of productivity of new oil discoveries in the United States for the familiar periods of 1965–71 and that of 1971–74 has to be analyzed. This trend should be examined to determine whether or not it would correspond to the crisis period.

The productivity of new oil discovery is usually shown by the number of barrels of newly discovered oil per foot of drilling activity during a specific year. To be able to come up with the corresponding figures for each year, however, we need to separate the volume of oil exploration from natural gas and other associated hydrocarbons. In this connection, Table 16 readily demonstrates the corresponding average productivity figures in the period of 1965–75 for the United States. The discovery rate for the period of 1972–75 shows a 15 per cent improvement over the period of 1966–71. Obviously, the above improvement is exactly the reverse of what has happened in the aged oil fields, and the price structure in the oil industry.

During the period of 1965–71 and that of 1971–74, the corresponding trends of productivity of new oil exploration exhibit 55 per cent decline and 113 per cent increase, respectively. Even if one disregards the extreme ly low productivity figure of 1971 and instead considers the periods of 1965–72 and 1972–74 for the analysis, one will come up with 23 per cent decline and 24 per cent increase, respectively. In addition, the corresponding average oil finding ratios for the periods of 1965–72 and 1972–74 are 14.27 barrel per foot and 15.39 barrel per foot.

As for the total finding rates (oil and gas) in the periods of 1966–71 and 1971–74, one finds a similar trend for both. More specifically, we have a 43 per cent decline in the productivity of hydrocarbon exploration in the pre-crisis and a nearly 39 per cent increase in the period which partly coincides with the crisis time span. In this connection, Table 17 shows the trends of productivity associated with the sphere of exploration for both pre-crisis and the crisis periods.

Table 17

Trends of the U.S. New Oil and Total Hydrocarbon Discovery Per Foot

Productivity Trend	*1965–71*	*1965–72*	*Periods* *1966–71*	*71–74*	*72–74*
Trend of New Oil Discovery	–55%	–23%	–42.2%	+ 113%	+24%
Trend of Total Hydrocarbon Discovery	–	–	–43.1%	+38.8%	+38.4%

Source: API, *Basic Petroleum Data Book,* and the U.S. *Department of Energy* Data.

What is significant here is that during the period which resulted in the oil crisis of 1973–74, productivity of both new oil and total hydrocarbon exploration increased substantially. Whereas, if the oil crisis had emerged from within the sphere of exploration, we would have had a decrease rather than an increase in the trend of that period. During the pre-crisis period, the productivity of exploration exhibits a tremendous decline. Moreover, this decline during the 1965-71 period did not have any significant influence upon the value and the price structure, since there was no fundamental price change. This is true for the period of 1971–74 as well. In other words, the condition of capital investment in the sphere of exploration does not conform to the condition of the oil crisis of 1973--74.[6] What is most likely in this respect is that the conditions for the emergence of the oil crisis should have actually evolved from other remaining spheres. These spheres, as we have pointed out, are the extension of exploration in the old oil fields and the intensification of production by way of enhanced recovery methods.

7. *Major Producing States, Depth Category and the General Trend of Oil Recovery*

In the present section, an attempt will be made to determine how and in what ways the recovery trends of oil fields in the major producing states, identified in Table 11, actually conform to that of the U.S. average oil recovery per well during the oil crisis.

In this respect, the oil field depth classification devised by the U.S. Bureau of Mines is used to identify the conditions of production prevalent in the oil fields of identical depth. There are five different depth categories, starting with (0–2000) and ending up with (8000–and deeper). In addition to the average oil recovery of the individual states, the size of the average oil recovery of different depth intervals associated with the oil fields of these states is also a factor that deserves attention. Table 18 contains the details of oil recovery figures associated with the major oil producing states.

Table 18

Average Oil Recovery of the U.S. Oil Fields in the Major Producing States According to Depth Classification

State	*Depth Categories (in feet)*	*Average Oil Recovery of Oil Fields (barrel/day)* 1965	1971	1974
		1965	*1971*	*1974*
Texas	0–2000	3.5	3.96	3.88
	2000–4000	7.99	10.17	10.02
	4000–6000	16.16	29.75	35.81
	6000–8000	25.60	36.12	43.23
	8000 and deeper	37.80	50.57	43.55
Louisiana	0–2000	2.96	7.97	7.17
	2000–4000	20.85	15.90	24.27
	4000–6000	34.10	11.44	15.53
	6000–8000	65.14	115.18	105.58
	8000 and deeper	89.74	132.91	140.00

(Table 18 cont.)

State	*Depth Categories (in feet)*	*Average Oil Recovery of Oil Fields (barrel/day)* *1965*	*1971*	*1974*
California	0–2000	10.22	13.63	12.83
	2000–4000	21.74	19.16	18.73
	4000–6000	26.15	39.80	36.66
	6000–8000	48.36	49.50	66.23
	8000 and deeper	43.83	65.00	39.52
Oklahoma	0–2000	2.41	4.55	4.73
	2000–4000	6.40	6.52	5.82
	4000–6000	9.78	13.01	10.23
	6000–8000	13.76	11.60	8.61
	8000 and deeper	21.63	8.21	17.47
Wyoming	0–2000	18.43	54.93	20.22
	2000–4000	66.75	41.66	52.12
	4000–6000	61.46	52.62	44.72
	6000–8000	48.35	58.24	56.13
	8000 and deeper	123.49	100.00	82.68
New Mexico	0–2000	11.14	7.82	6.15
	2000–4000	11.56	12.10	12.90
	4000–6000	19.20	22.25	16.16
	6000–8000	26.76	32.00	40.05
	8000 and deeper	70.93	54.30	38.38
Kansas	0–2000	1.86	3.91	1.43
	2000–4000	7.83	6.28	4.74
	4000–6000	15.12	14.36	9.32
	6000–8000	44.52	22.77	8.83
	8000 and deeper	0	0	0
Mississippi	0–2000	0	149.77	0
	2000–4000	28.30	40.54	38.16
	4000–6000	32.90	40.54	38.31
	6000–8000	58.79	36.44	29.74
	8000 and deeper	65.24	70.81	61.62

Source: Compiled by the author from:

U.S. Department of the Interior, 'Depth and Producing Rate Classification of Oil Reservoirs in the 14 Principal Oil Producing States', *Bureau of Mines Information Circular,* 8362 (Washington, D.C.: Government Printing Office, 1967);
U.S. Department of the Interior, 'Depth and Producing Rate Classification of Petroleum Reservoirs in the United States', *Bureau of Mines Information Circular,* 8675A (Washington, D.C.: Government Printing Office, 1976); and
U.S. Department of Energy, *Depth and Producing Rate Classification of Domestic Oil Reservoirs, 1974* (Washington, D.C.: Government Printing Office, 1978).

Table 19

*Changes in the Average Oil Recovery Per Well**
1965-1974

State	Depth Categories (in feet)	1965-71	1971-74
Texas	8000 and deeper	+0.34	-0.14
Louisiana	6000-8000	+0.77	-0.08
California	8000 and deeper	+0.48	-0.38
	4000-6000	+0.54	-0.075
Oklahoma	4000-6000	+0.30	-0.23
Wyoming	6000-8000	+0.21	-0.034
	0-2000	+2.05	-0.64
New Mexico	4000-6000	+0.16	-0.27
Kansas	0-2000	+1.05	-0.64
Mississippi	8000 and deeper	+ .015	-0.13
	2000-4000	+0.46	-0.07
	0-2000	—	-1.0
Oklahoma	6000-8000	-0.16	-0.26
Wyoming	8000 and deeper	-0.19	-0.17
	4000-6000	-0.14	-0.15
New Mexico	8000 and deeper	-0.23	-0.29
	0-2000	-0.30	-0.21
Kansas	6000-8000	-0.49	-0.61
	4000-6000	-0.05	-0.35
	2000-4000	-0.20	-0.24
Mississippi	6000-8000	-0.38	-0.18

* Figures are in percentages.

Source: Compiled by author from the same three main sources as used for Table 18.

Due to the existence of different geological formations in the oil producing states, no single depth category by itself can reflect the average conditions. The conditions of capital investment made on the oil fields are not identical but vary from one field to another. Therefore, the conditions of oil recovery and productivity of any two oil fields that are located in a particular state, or any two oil fields of particular depth category that belong to different producing states, do not seem to be similar.

Referring to the clearest case, however, the oil fields located in Texas (8000-and deeper), Louisiana (6000-8000), California (8000-and deeper, and 4000-6000), Oklahoma (4000-6000), Wyoming (6000-8000, and 0-2000), New Mexico (4000-6000), Kansas (0-2000), and Mississippi (8000-and deeper, 2000-4000, and 0-2000) show a considerable decline of recovery per well during the period of 1971-74. The above oil fields are all ones which exhibited oil recovery increase during the period of 1965--71.

There are some depth categories that associate with decline during both periods. These are: Oklahoma (6000-8000), Wyoming (8000-and deeper, and 4000-6000), New Mexico (8000-and deeper, and 0-2000), Kansas (6000-8000, 4000-6000, and 2000-4000), and Mississippi (6000-8000).[7] (See Table 19.)

It was pointed out earlier that when the impact of MDF is considered in the Texas and Louisiana oil fields, one must revise the oil recovery figures for 1965 and 1971. This revision would ultimately change the trend of oil recovery per well in the related depth brackets of these particular states for the periods of 1965-71 and 1971-74. Likewise, the upward revision due to MDF would produce a downward trend in the oil recovery of these states for the period of 1965-71.[8] In this case, oil fields in Texas (8000-and deeper) and Louisiana (6000-8000) should be removed from the clearcut case and placed under an ambiguous category.

8. The Role of Productivity Decline in the Oil Crisis of 1973-74

Throughout the previous sections, we tried to define the notion of average oil recovery of an oil field, to specify the major producing states, to identify the depth categories of the oil fields, to distinguish between the crisis and non-crisis periods, and to measure the changes that occurred during such periods. The remaining task, now is to establish the relationship between the prevailing state of oil recovery and the crisis.

The trend of the average of oil recovery is a clear reflection of evolving production conditions prevailing in the oil fields. Given the geological structure of the oil reservoirs, the production conditions are the consequence of the way in which capital investments are made for reserve production in the entire industry. The concentration of capital investments in the old oil fields, if widespread, can require the restructuring of the entire industry. In other words, if the cost structures of production in the old oil fields, due to further 'capital deepening' were to rise, then the magnitude of individual value and possibly the oil value at the industry level has to change as well.

It turns out that the trend of the average U.S. oil recovery, i.e., the trend of the weighted average of daily barrels of oil per well, for the period of 1965-71, shows a considerable increase. However, the same trend for the period of 1971-74 i.e., the period that partly coincides with the crisis does indicate a tremendous decline (see Table 12). Considering that investments are either done in the form of 'capital widening' or by way of 'capital deepening', the *extension* of the existing oil fields or the massive applications of enhanced recovery methods would necessarily require capital investment in the form of 'capital deepening'. The example of 'capital widening' is the investment in the sphere of new oil discovery which actually resulted in a decline in the average finding rate in the period of 1965-71 and an increase during the 1971-74 period. If, in fact, production in such a sphere had something to do with the crisis, one would have expected at least a decline in the average finding rate during the period of 1971-74.[9]

In addition to the above conditions, the average life span of U.S. oil wells was substantially shortened during the period of 1971-74, as compared with the trend of 1965-71.[10] This shows that the depletion of existing oil fields increased substantially as further capital investments were made on the old reservoirs on a massive scale during the 1971-74 period. Meanwhile, the rate of cumulative abandonment of oil wells did actually decline during the 1971-74 period. This is the point at which the producing oil wells are considered to be exhausted and therefore are abandoned. As Table 20 indicates, the average rate of oil well abandonment per year for 1971-74 is 1.54 percent, which is half the size of corresponding figure for the 1965-71 period. Consequently, one has to conclude that in the period of 1971-74 a substantial number of oil wells, which would have been abandoned in the previous period, were retained for further production. Obviously in order to keep the production from such oil wells flowing, additional investments of capital were necessary. Given the conditions of the oil wells, the average productivity of such investments has to have declined during the period of crisis.

Table 20

U.S. Oil Well Abandonment and the Life Span of Oil Wells (1965-1974)

Subject	*1965-71*	*1971-74*
The Life Span of Oil Wells (years)	26.014	24.25
Cumulative Abandonment Rate of Oil Wells	+21.38%	+ 6.18%
Cumulative Abandonment Rate of Oil Wells Per Year	+ 2.05%	+ 1.54%
Average Number of Oil Wells Abandoned Per Year	24,749 (wells)	17,187 (wells)

Source: Compiled by the author from API, *Basic Petroleum Data Book*, 1979.

As we have pointed out earlier, the oil crisis of 1973-74 is the consequence of the generalization of the conditions of production in the old oil fields of the U.S. for the entire industry. At the same time, due to the existence of a particular property relation within the oil industry, the least productive firm tends to determine the magnitude of value for the entire industry. The relations of oil property take the phenomenal form of oil rent and royalties. Consequently, the more productive regions, in addition to their share of *normal profit,* appropriate a *surplus profit* in the form of oil rents or royalties.

Thus we conclude, the generalization of the condition of production in the least productive U.S. oil fields led to the crisis, which in turn established a new basis for the entire oil industry. Given the extent of internationalization of production, this new basis constituted the formation of new value, oil rent, and market prices for the entire oil industry.

NOTES

1. U.S. Department of the Interior, 'Depth and Producing Rate Classification of Oil Reservoirs in the 14 Principal Oil Producing States', *Bureau of Mines Information Circular* 8362 (Washington, D.C.: Government Printing Office, 1967); U.S. Department of the Interior, 'Depth and Producing Rate Classification of Petroleum Reservoirs in the United States', *Bureau of Mines Information Circular* 8675A (Washington, D.C.: Government Printing Office, 1976); U.S. Department of Energy, *Depth and Producing Rate Classification of Domestic Oil Reservoirs, 1974* (Washington, D.C.: Government Printing Office, 1978).
2. In 1965, Alaska was not considered an oil state. Instead, Illinois was among the top nine oil producing states.
3. See John Blair, *Control of Oil* (New York: Pantheon Books, 1976), pp. 159-63.
4. It is important to note that we are concerned with the average oil recovery per well associated with the *entire round of capital investment* applied to the oil fields rather than the so-called investment at the margin. See Ben Fine, 'On Marx's Theory of Agricultural Rent', *Economy and Society,* August 1979, pp. 257-58.
5. The period of 1971-74 is merely selected for the sake of consistency with other sections of the present chapter.
6. It should be understood that the oil crisis of 1973-74, contrary to the general periodic crises, evolved through the generalization of a process which prevailed in less productive regions, rather than the generalization of new and more productive techniques.
7. The ambiguous category of the oil fields is the one which exhibits either a decreasing trend or increasing trend over *both* 1965-71 and 1971-74.
8. Market Demand Factor, which is a form of production restriction, was dominant in Texas and Louisiana prior to 1972.
9. It should be noted that we are not talking about the necessity or lack of necessity of fresh reserves. What we have in mind is the fact that in the final analysis production conditions in the old oil fields were dominant in the actual formation of values and prices.
10. See API, *Basic Petroleum Data Book,* 1979.

CHAPTER VIII

THE ECONOMETRIC MODELS OF OIL PRICES AND OIL RENTS

1. Introduction

The present chapter deals with the formulation of econometric models of long-run United States oil prices, as well as models of oil rents associated with the United States and the Middle East oil regions.

In this connection, we will attempt to demonstrate that the increase in the U.S. oil development capital expenditures has a high degree of association with the increase in the price of oil during the 1973–74 crisis. In addition, it will be shown that the increase in the amount of differential oil rent was the consequence of the increase in the cost of long-run investments associated with the old U.S. oil fields.

The analysis is divided into six sections. In section 2 an attempt will be made to state the theoretical underpinnings of our analysis, both in relation to the formation of new market value as well as the formation of a higher level of differential oil rent in the oil industry. In section 3 the statements regarding the objectives of the econometric models will be presented. Section 4 is the place where the nature and characteristics of statistical data are discussed. The method of analysis and a description of techniques are introduced in section 5. Section 6 will contain the econometric models of oil prices and oil rents. Finally, in section 7 the summary of major points and a conclusion will be presented.

2. The Theoretical Underpinnings

It has been argued that the oil crisis of 1973–74 is the symptom of an increase in the magnitude of individual value produced in association with the old U.S. oil fields. At the same time, being associated with the least productive region of the world, this individual value has become the industry's regulating *value.* Due to the competition with the individual values produced from various oil regions, the least productive region established the magnitude of the industry value. Thus, the oil crisis is the consequence of the negation of the old value and the emergence of the newly established value for the entire oil industry.

Oil producers in the more productive regions are able to appropriate a portion of surplus value produced within the industry in the form of *differential oil rent,* the size of which depends upon the difference in the magnitude of the industry value and the individual values of the more productive oil regions. As the old value is replaced by the newly established industry value, the difference between the magnitudes of individual values and that of the industry value increases. In other words, the increase in the magnitude of market value at the industry level has led to the increase in the size of differential oil rent developed in all the oil regions. The largest increase in differential oil rent has been associated with the producers in the most productive oil regions.

3. The Objectives of Econometric Models

One way to account for the relationship between the long-term trend of the U.S. oil prices and that of the total cost of investments is through the formation of linear econometric models. These models are constructed to accommodate the relationships of oil prices and the capital investments associated with different spheres of oil production, such as new exploration and production from the aged U.S. oil fields. It is also possible to establish a relationship between the price of oil, on the one hand, and the long run supply and demand conditions in the U.S. crude oil market, on the other. Such a relationship, if not significant, suggests that the shifts in demand and supply conditions, while important in the short run, do not seem to influence the U.S. oil prices in the long run.

In order to account for the phenomenon of differential oil rent, which results from the existence of oil property relation, on the one hand, and *competition* among the individual values produced on different oil regions of the world, on the other, the econometric models of oil rents for the Middle East and the United States, i.e., the most and the least productive oil producing regions of the world, are also formulated.

There are a total of four econometric models. These include two linear models for the long-run price of oil and two linear models for the Middle East and the U.S. oil rents. The first two price models show the association of capital costs to prices, while the last two relate the oil investments to oil rents for both the Middle East and the United States. In order to account for the difference between the pre-crisis and the crisis periods, appropriate dummy variables are included in the models.

4. The Source and Characteristics of Statistical Data

The statistical data utilized for the present chapter are related to various aspects of the oil production both in the United States and the Middle East. The U.S. data are used to estimate the United States long-run exploration, development and total capital costs of oil production during the period of 1947–75. Also, we aimed to calculate the long-run oil investments in the Middle East from the available Middle Eastern sources.

Table 21 shows the calculated trends of the U.S. oil exploration costs, development costs, and total capital costs per barrel. It also includes the daily excess demand for oil and the U.S. oil prices for the period of 1947–75. Table 22 presents the time series of total U.S.. capital costs, total Middle East (M.E.) capital costs, and the oil royalties for these two regions.[1] (The sources of the above data are the American Petroleum Institute and the U.S. Department of Energy.)

5. *The Method of Analysis and Description of Techniques*

In previous sections, the theoretical basis and the purpose of econometric models were discussed. Our task in the present section is to describe the characteristic of specific statistical methods and techniques that are used in this chapter.

The method of multiple regression (OLS) is used for the estimation of the coefficients of the models. The period of analysis covers some twenty-nine years of post-World War II oil development. At the same time, a distinction has been made between the pre- and the crisis periods by accommodation of appropriate dummy variables in the equations. The advantage of such a technique is that it can reveal the significance (or the lack of significance) of changes in the equations.

There are different versions of dummy variable techniques and each can be implemented according to the scope of a particular problem. The widely used version however is the *additive* form of dummy variable. Often the aim of this form is to account for the probable shifts of the intercep in the regression models, or the qualitative changes that might have taken place outside these models. If there is a need for determination of anticipated changes in the magnitude of a specific independent variable in a regression equation, however, one has to construct a column of dummy variables which has zero values for one segment and the actual values of the same variable in question for the remaining part of the period. This type of dummy variable is referred to as *multiplicative* in the econometrics literature.[2] The usage of such a variable in the equation can measure the extent of the probable changes that might have taken place in connection with the independent variable of the model. Since in this chapter,we are interested in the effect of the changes of the U.S. long-run capital costs on the oil prices, the latter type of dummy variables are used in the price models. However, we used the additive type of dummy variable in the oil rent models in order to account for the probable changes in the intercepts.

6. *The Econometric Models of the Long-Run Oil Prices and the Oil Rents*

There are four different models developed in the present section. The first model explains the relationship of the oil price and the long-run development expenditures per barrel made in the United States. The second model focuses on the relationship of the price of oil and long-run total capital costs of oil in the United States. The two remaining models, however, are about the relationship of capital investments and the oil royalties in the Middle East and the United States. These models are studied under separate headings of long-run oil prices and the oil rents for the sake of simplicity.

i.. The Econometric Models of the Long-Run United States Oil Prices

The primary purpose for the construction of the price models is to be able to account for the relationship of the oil price and capital costs in order to focus on the changes that have taken place during the oil crisis of 1973–74. Furthermore, these models can reflect the nature of the changes in the cost structure that preceded the changes of the price structure during the outbreak of the crisis in the oil industry as a whole.

These models are:

$$(1)\quad P = \alpha_0 + \alpha_{,1} C^d + \alpha_2 C^e + \alpha DC^d + \mu$$

$$(2)\quad P = \beta_0 + \beta_1 C^t + \beta_2 X + \beta_3 DC^t + e\,,$$

where P is the wellhead price of oil in the United States, C^d is the long-run development expenditures per barrel, C^e is the long-run exploration expenditures per barrel, DC^d is the multiplicative type of dummy variable measuring the probable change in the magnitude of development investment per barrel, C^t is the long-run total U.S. capital costs per barrel, X is the daily excess oil demand in the U.S. market, and DC^t is the multiplicative form of dummy variable used in order to detect the probable changes that might have occurred in the relationship of prices and total capital costs since the oil crisis. Finally 'μ' and 'e' respectively are the disturbance terms of the first and the second equations.

Before presenting our the empirical results, we need to make three additional remarks. These are: (1) the importance of the distinction between the oil development costs (C^d) and the oil exploration costs (C^e); (2) a clarification about the notion of *excess demand* in the U.S. oil market and its relationship with the oil prices; and (3) the usage of the multiplicative type of dummy variables in the above models and the significance of the separation of pre-crisis and the crisis periods.

First of all, the category of oil development cost refers to the type of capital expenditures that are invested in the previously discovered oil fields for the purpose of additional oil recovery. In contrast, the oil exploration expenditures are the type of capital costs made for new exploration or new oil discovery. The first category is also synonymous with investments for 'extensions and revisions', technical jargon often used, in the oil industry. In other words, from the standpoint of economic theory, the exploration and development costs are respectively synonymous with the notions of 'capital widening' and 'capital deepening'.

The second issue is *excess demand,* an independent variable used in the second equation. The notion of excess demand is necessary for a description of the long-run demand and supply conditions in the U.S. oil market. The amount of the U.S. oil refinery input is considered to be the demand for domestic crude oil, while the total quantity of domestic oil production plus the amount of crude oil imported are considered as total supply, providing the oil stocks (inventories) do not change. The differences between the quantity demanded and quantity supplied therefore result in either excess demand or excess supply.[3] As Table 21 shows, during the period of 1947–67, except for only three random years, the U.S. oil market was over supplied. However, the period of 1968–75 is the period of the existence of excess demand in the U.S. oil market.

Finally, in order to separate the two sub-periods of pre-crisis, we devised appropriate dummy variables in the equations. The entire period of study, i.e., 1947–75, is divided into the sub-periods of 1947–71 and 1972–75 in order to make a comparison between the magnitudes of capital cost coefficients. The dummy variables used are of multiplicative form. This type of dummy variable is the only one which would account for the probable changes in the coefficients of explanatory variables rather than the change in the intercept. The first equation, i.e., $P = \alpha_0 + \alpha_1 C^d + \alpha_2 C^e + \alpha_3 DC^d + \mu$ includes DC^d, or the dummy variable that accounts for the change in C^d's coefficient. DC^t is the dummy variable associated with the change of C^t in the second equation. Tables 21 and 22 show the statistical data associated with these variables. The statistical technique used to fit the data to the above models is ordinary last square (OLS).

The regression models for the estimation of the parameters of price equations are:

Equation I: $$\hat{P} = 1.56957 + \underset{(0.46269)}{1.213826C^d} - \underset{(0.18099)}{0.095069C^e} + \underset{(0.34145)}{0.5853917DC^d}$$

$n = 29$

$R^2 = 0.87505$

Equation II: $$\hat{P} = 1.026874 + \underset{(0.44236)}{1.388215C^t} - \underset{(0.79733)}{1.338945X} + \underset{(0.31542)}{0.605574DC^t}$$

$n = 29$

$R^2 = 0.86594$

Table 21

The U.S. Prices, Capital Costs, and Daily Excess Demand of Crude Oil 1947-1975

Year	The U.S. Wellhead Price of Oil	Total U.S. Capital Costs of Oil Per Barrel*	The U.S. Oil Development Costs Per Barrel	The U.S. Oil Exploration Costs Per Barrel	Daily U.S. Excess Demand For Oil (Millions of Barrels)
1947	1.93	0.668	0.664	0.684	-0.279
1948	2.60	0.537	0.491	0.930	-0.276
1949	2.54	0.600	0.653	0.464	-0.230
1950	2.51	0.793	0.793	0.791	-0.155
1951	2.53	0.587	0.486	1.630	-0.154
1952	2.53	1.147	1.058	1.549	-0.159
1953	2.68	1.042	0.966	1.387	-0.106
1954	2.78	1.308	1.290	1.379	-0.040
1955	2.77	1.428	1.331	1.917	-0.109
1956	2.79	1.486	1.342	2.262	-0.148
1957	3.09	1.808	1.685	2.420	-0.273
1958	3.01	1.338	1.167	2.584	-0.057
1959	2.90	0.985	0.857	2.126	-0.025
1960	2.88	1.442	1.307	2.566	+0.017
1961	2.89	1.200	1.133	1.623	-0.044
1962	2.90	1.629	1.607	1.731	-0.048
1963	2.89	1.546	1.457	2.008	+0.014
1964	2.88	1.351	1.221	2.220	-0.005
1965	2.86	1.106	1.048	1.421	+0.001
1966	2.88	1.138	0.984	2.456	-0.076
1967	2.92	1.190	1.048	2.270	-0.124
1968	2.94	1.770	1.579	2.886	+0.362
1969	3.09	1.929	1.646	4.072	+0.443

Table 21 (continued)

Year	*The U.S. Wellhead Price of Oil*	*Total U.S. Capital Costs of Oil Per Barrel**	*The U.S. Oil Development Costs Per Barrel*	*The U.S. Oil Expoloration Costs Per Barrel*	*Daily U.S. Excess Demand For Oil (Millions of Barrels)*
1970	3.18	1.193	1.055	2.209	+0.366
1971	3.39	1.272	1.075	3.987	+0.486
1972	3.39	2.897	2.787	3.404	+0.482
1973	3.89	2.308	2.068	4.590	+0.403
1974	10.13	3.953	3.885	4.321	+0.281
1975	12.03	5.319	5.769	3.797	+0.330

* A Barrel is 42 U.S. Gallons.

Source: API, *Reserves of Crude Oil, Natural Gas Liquids, and Natural Gas in the United States and Canada as of December 31, 1979*, vol. 34, (June, 1980), p. 24, Table II; API, *Basic Petroleum Data Book*, vol. 1, no. 1, section V, Table 9, Section III, Table 2; and DOE, *Annual Report to Congress*, vol. 2, 1979, and 1981.

The figures in parentheses are standard errors of estimation. In the equation I, variable C^e, i.e., the exploration costs per barrel, is not statistically significant in relation to the price trend during the entire period of 1947–75. In equation II, variable X does not seem to be statistically significant in relation to the oil price. Also, both DC^d and DC^t—dummy variables associated with both of the above equations—are *not* statistically significant. The level of significance chosen for the above tests is 95%. The regression results associated with the first and second equations indicate that both C^d and C^t are significantly related to the trend of the U.S. oil prices. These variables stand for the U.S. oil development costs per barrel and the U.S. total capital costs per barrel during the entire period of 1947–75.

From equation I, we may conclude that (1) the long-term development costs or the investments in the sphere of old oil fields are, statistically speaking, significantly related to the U.S. oil prices during the entire period of 1947–75, (2) the investments in the area of new exploration do not have any significant relationship with the trend of the oil prices, and (3) since DC^d is not statistically significant, one has to conclude that there is no significant structural divergence between the pre-crisis and the post-crisis periods. This implies that the changes of the price structure are *closely* related to the changes in the level of development capital costs during the crisis period. (See Figure 3.) At the same time, we encounter neither the problem of autocorrelation nor of multicollinearity in both models. (See Appendix C.)

Equation II as we have already explained, is the relationship between the total long-run oil costs and the oil prices. The only significant variable in the regression equation is C^t. From the equation II, we may conclude that (1) the total capital costs are significantly related to the U.S. oil prices during the entire period of 1947–75, (2) the supply and demand conditions do not have any significant relation with the long-run oil prices, even though this conclusion might not be true in the short run, and (3) since DC^t is not statistically significant, there seems to be no significant structural divergence during the entire period of 1947–75. As a result, it would be appropriate to conclude that the changes in the oil prices are significantly related to the changes of total capital costs during the crisis period. (See Figure 4.) Since the total costs trend is almost identical to that of the development costs, it is apparent that the same relationship has to be exhibited from both equations.

It should be understood that C^t and C^d are the long-run total capital and development investments per barrel, respectively. While C^t is the fundamental basis of the long-run U.S. oil prices, it will certainly differ from the U.S. total cost per barrel. The total cost per barrel includes the production or lifting cost per barrel, other operating costs per barrel, the cost of leases, taxes, etc. Because of this

Figure 3

The Actual and Predicted Trends of the U.S. Oil Prices During the Period of 1947–1975 Based on Development Costs

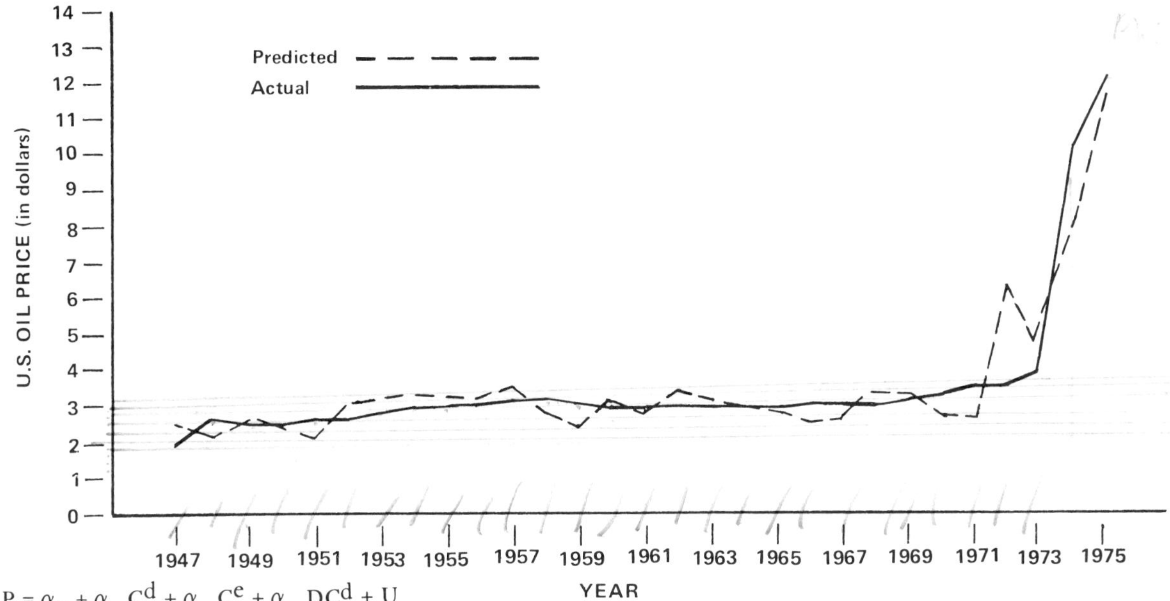

$P = \alpha_0 + \alpha_1 C^d + \alpha_2 C^e + \alpha_3 DC^d + U$

where P is the U.S. price of oil, C^d is the oil development costs per barrel, C^e is the exploration costs per barrel, and DC^d is the dummy variable used to account for the change in the C^d coefficient during the oil crisis.

Source: API, *The U.S. Reserves of Crude Oil, Natural Gas. . .*, API, 1981 and API, *Basic Petroleum Data Book*, various issues.

Figure 4

The Actual and Predicted Trends of the U.S. Oil Prices During the Period of 1947–1975, Based on Total Capital Costs

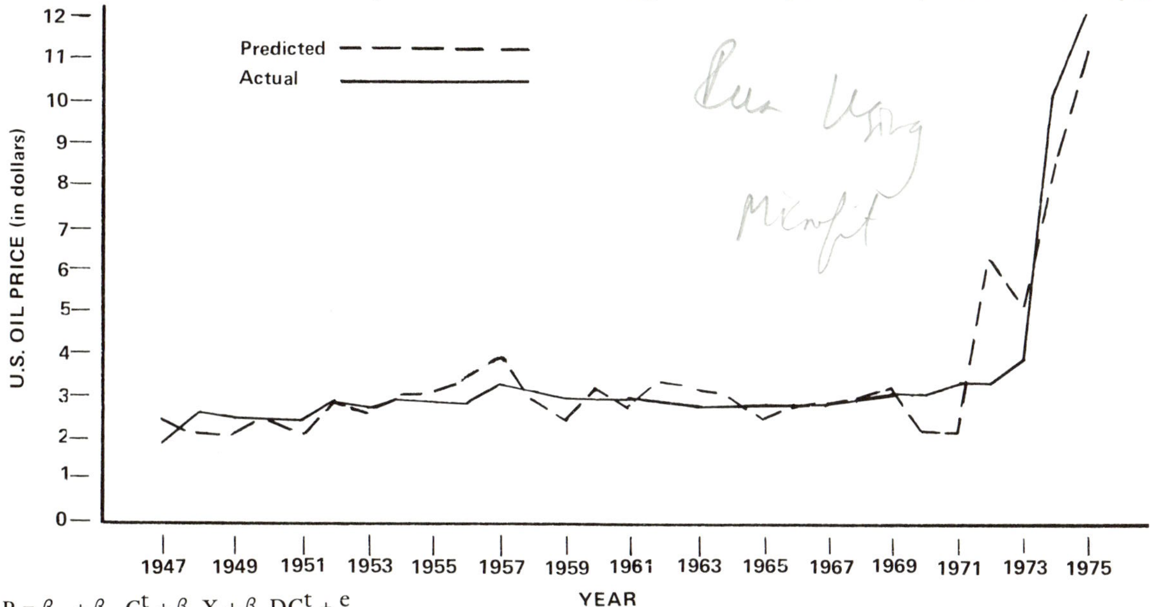

$$P = \beta_0 + \beta_1 C^t + \beta_2 X + \beta_3 DC^t + e,$$

where P is the U.S. price of oil, C^t is the total capital costs per barrel, X is the daily excess demand, and DC^t is the dummy variable accounting for the change in the coefficient of C^t, i.e., the slope of cost-price function, during the oil crisis.

Source: API, *The U.S. Reserves of Crude Oil, Natural Gas. . .*, API, 1981, and API, *Basic Petroleum Data Book*, various volumes.

difference one has to demonstrate that the long-term trends, total as well as development capital investments per barrel, are significantly associated with that of the long-run prices. The above models show that during the pre-crisis as well as the crisis periods, as a result of a unit increase in C^d and C^t, oil prices increase about 1.21 unit and 1.39 unit respectively.

The most important point in this chapter is the relationship between the drastic price change and the change in the long-run investment per barrel during the oil crisis period. One way to determine the above relationship is through the examination of the existence of structural conformity or divergence in these models. This can be done through the usage of multiplicative dummy variables in order to account for the changes in the slopes of both price-cost functions.[4]

Our primary purpose for development of these econometric models is to demonstrate (1) that there is a high degree of association between the trend of the wellhead price of the crude oil and that of the total capital expenditures per barrel in the United States during the period under study, (2) that the only cost category which has been significantly associated with the trend of the oil prices is the development capital cost, and (3) that the price increase was a response to restructuring of production and an increase in level of development capital costs which had already started in 1972.

In order to be able to interpret the above empirical findings, one has to make a distinction between the pre-crisis and crisis periods. After the inclusion of appropriate dummy variables in the above models and having discovered that their corresponding coefficients are not statistically significant, it turns out that even though we do not encounter any structural divergence during the period of 1947–75, nevertheless we are dealing with the two very dissimilar periods in the history of the U.S. oil production during the post-war period. But one has to realize that the material basis of such *historical* dissimilarities lies in the structural conformity of prices *vis-a-vis* the changing pattern of capital investments, that are shown *econometrically*. In this context, a close examination of data reveals that the total as well as development costs per barrel both started to rise earlier in 1972, whereas the oil price increase did not come about until late 1973. In other words, the price increase was lagging behind the cost increase prior to the crisis. It is through the crisis itself that the oil production is restructured and ultimately the new price level emerged as its corresponding outward manifestation.[5]

Consequently, the implication of the above analysis is to support the hypothesis that the long-run cost increases and the oil price increase of 1973–74 are related to each other. Theoretically, it is reasonable to maintain that the oil price increase was the result of the long-run increase in the U.S. development costs per barrel during the early 1970s. In addition, this type of investment is associated with the development of oil reserves from the previously discovered and aged U.S. oil fields located in the lower 48 states.

Therefore, from the evaluation of the above equations together we conclude that (1) the changes in the total costs of oil were significantly related to the oil price changes, (2) the sphere of exploration costs associated with the new oil discovery did not have any significant relation with the oil price increase, (3) the development costs associated with the oil recovery from the old oil fields are significantly related to the price trend, and (4) the *increase* in the oil development costs per barrel is significantly associated with the *increase* in the price of oil during the oil crisis.

To sum up, the cost structure of the old oil fields has been the basis of value formation and market prices during the entire period of 1947–75. In addition, as the capital investment per barrel in the above sphere of production increased, it led to the formation of a new value and ultimately new prices of higher magnitude.

ii. The Econometric Models of the United States and the Middle East Oil Rents

Our purpose for the formation of the following models is threefold: (1) to find the magnitude of oil rent within each individual oil region; (2) to observe the magnitude of differential oil rents associated with the least and the most productive oil regions of the world; and (3) to demonstrate that the increase in the differential oil rents was the consequence of the changes in the cost structure of the U.S. oil region, which simultaneously led to the change in the value and prices of oil at the international level.

The models are:

$$R_{M.E.} = a_0 + a_1 C_{M.E.} + D_1 + e$$

$$R_{U.S.} = b_0 + b_1 C_{U.S.} + D_2 + v,$$

where $R_{M.E.}$ is the total receipt per barrel of the oil-exporting governments of the Middle East, $C_{M.E.}$ is the total capital cost per barrel in the Middle East oil region, D_1 and D_2 are dummy variable used to account for the change in the intercept of each model from the time of the initiation of \$11.65 as a 'posted price' by OPEC and the establishment of \$10.13 per barrel for the wellhead price of oil in the United States (1974), $R_{U.S.}$ is the total receipt per barrel of oil by the land owners in the United States and $C_{U.S.}$ is the total capital cost per barrel of oil in the U.S. (parameters 'e' and 'v' are the residuals of the equations).

It should be pointed out that the distinction between the pre-crisis and post-crisis periods for the oil rent equations is based on the time that 'posted price' reached the level of $11.65 (Jan. 1, 1974). There arrangement of the sub-periods in this section is different fr previous one. Thus, we attempt to divide the period of 1950–78 i sub-periods of 1950–73 and 1974–78. The designation of slightly dif periods of analysis for the present section is due to the lack of adequate data for the Middle East oil region during the first few years.

If the data in Table 22 are fitted into the above models, the parameters of the regression equations can be estimated as follows:

$$\text{Equation III:}\quad \hat{R}_{M.E.} = \underset{(10.31118535)}{56.74344822} + \underset{(1.45512002)}{9.09708682\,C_{m.e.}} + \underset{(30.14084588)}{832.70879405\,D_1} \qquad n = 29,\quad R^2 = 0.985938$$

$$\text{Equation IV:}\quad \hat{R}_{U.S.} = \underset{(0.02529289)}{0.32459884} + \underset{(0.01484322)}{0.08454925\,C_{U.S.}} + \underset{(0.08582918)}{0.96271473\,D_2} \qquad n = 29,\quad R^2 = 0.985334$$

The figures in parentheses are standard errors of estimates, and they all exhibit that the explanatory variables in both equations are 99 per cent statistically significant. There is no problem of multicollinearity. However, there is a problem concerning autocorrelation in equation IV, i.e., the U.S. oil rent model. Therefore, the attempt has been made to solve the problem of autoregressive disturbances and find the corrected value for the coefficients of the U.S. oil rent model. (See Appendix C.) Since the rent and the cost associated with the Middle East oil rent model are in U.S. cent units, we divided the results by 100 in order to denominate them in U.S. dollars. The revised models are:

$$\text{Equation V:}\quad \hat{R}_{M.E.} = \underset{(0.10311185)}{0.56743448} + \underset{(1.45512002)}{9.09708682\,C_{M.E.}} + \underset{(0.30140846)}{8.32708794\,D_1} \qquad n = 29,\quad R^2 = 0.985938$$

$$\text{Equation VI:}\quad \hat{R}_{U.S.} = \underset{(0.01825457)}{0.31110493} + \underset{(0.01154445)}{0.09282300\,C_{U.S.}} + \underset{(0.06616019)}{0.92433064\,D_2} \qquad n = 29,\quad R^2 = 0.985334$$

Table 22

*The Middle East and the U.S. Capital Expenditures and Oil Rents per Barrel**
1950–1978

(Figures in U.S. Dollars)

Year	U.S. Oil Royalty Per Barrel (15% of oil price)	Middle East Oil Receipts Per Barrel	Total U.S. Oil Capital Costs Per Barrel	Total M.E. Oil Capital Costs Per Barrel
1950	0.376	0.370	0.793	0.0082
1951	0.379	0.410	0.587	0.0072
1952	0.379	0.530	1.147	0.0091
1953	0.402	0.660	1.042	0.0070
1954	0.417	0.720	1.308	0.0054
1955	0.415	0.760	1.428	0.0047
1956	0.418	0.780	1.486	0.0105
1957	0.463	0.800	1.808	0.0085
1958	0.451	0.790	1.338	0.0538
1959	0.435	0.790	0.985	0.0354
1960	0.432	0.752	1.442	0.0795
1961	0.433	0.735	1.200	0.0430
1962	0.435	0.722	1.629	0.0374
1963	0.433	0.790	1.546	0.0113
1964	0.432	0.762	1.351	0.0300
1965	0.429	0.785	1.106	0.0710
1966	0.432	0.800	1.138	0.0130
1967	0.438	0.850	1.190	0.0200
1968	0.441	0.882	1.770	0.0140
1969	0.463	0.860	1.929	0.0600
1970	0.477	0.895	1.193	0.0110
1971	0.508	1.205	1.272	0.0270
1972	0.508	1.302	2.897	0.0390
1973	0.660	2.000	2.308	0.0900
1974	1.519	7.825	3.953	0.0570
1975	1.804	9.820	5.319	0.0470
1976	1.824	10.610	8.729	0.0960
1977	2.038	12.155	7.343	0.3870
1978	2.092	11.595	8.251	0.2410

* See Appendix D for calculations.

Source: API, *Reserve of Crude Oil, Natural Gas Liquids, and Natural Gas in the U.S. and Canada as of December 31, 1979,* vol. 34, June, 1980, p. 24, Table II; Department of Energy, *Annual Report to Congress,* vol. 2, p. 39, Table 18; API, *Basic Petroleum Data Book,* vol. 1, no. 1, 1981, Section IV, Table 9; *Petroleum Outlook,* May, 1976 and 1979; Frank R. Wyant, *The United States, OPEC, and Multinational Oil* (Lexington, Mass.: D.C. Heath & Company, 1979), p. 169, Table A-2; C. Issawi and M. Yeganeh, *The Economics of Middle Eastern Oil* (New York: Praeger Publishers, 1962), p. 114, Table 34.

The comparison of the above equations reveals that (1) the Middle East oil region is *absolutely* more productive than the U.S. oil region, (2) following the oil crisis the level of differential oil rents, both within the United States and the Middle East has increased, (3) the volume of differential oil rents within the international oil industry has increased following the crisis, and (4) the amount of differential oil rents associated with the Middle East oil region, i.e., the most productive oil region of the world, increased more than any other area. (See Figures 5 and 6.)

The first point can be proved from the comparison of the coefficients of $C_{M.E.}$ and $C_{U.S.}$ in the above equations. This comparison shows that for each unit of capital investments there is more oil discovery and therefore higher oil rent in the Middle East *vis-a-vis* the U.S. oil region. The second point can be substantiated by the changes of the intercepts of both models, that are captured by dummy variables. In the U.S. model the differential change of intercept since 1974 is about $0.92 per barrel. By the same token, the differential change of the intercept of the Middle East model exhibits an addition of $8.33 per barrel. The third point follows from our previous arguments. If the level of differential oil rent has increased in the least as well as the most costly oil regions, then it seems reasonable to imply that it must have increased for all the regions. The fourth point deals with the magnitude of the increase in differential oil rents. The examination of the size of the differential intercept associated with the Middle East oil rent model indicates that the producers in this region are able to appropriate the highest level of differential oil rent in the oil industry.

7. *Conclusion*

The oil crisis of 1973-74 is the result of the reorganization of oil production and the formation of new value and prices at the international level. The unprecedented increase in the production costs of oil from the old U.S. oil fields, on the one hand, and the formation of new industry value upon these fields have led to a price increase at the international level. At the same time, the formation of new value at the industry resulted from competition among individual values produced in various oil regions of the world. Since the conditions of production in the least productive region determined the magnitude of value, the producers of more productive oil regions are able to appropriate a portion of *surplus value* in the form of differential oil rent.

Our conclusion about the price models is that the U.S. oil price has a significant association with the total capital costs and development costs per barrel. At the same time, the changes in the oil prices are significantly associated with the changes in the total capital costs and development costs per barrel during the oil crisis of 1973-74. Statistically speaking, we found no significant relationship between the exploration costs per barrel

Figure 5

The Actual and Predicted Trends of the Oil Rent per Barrel in the Middle East Oil Region, 1950–1978

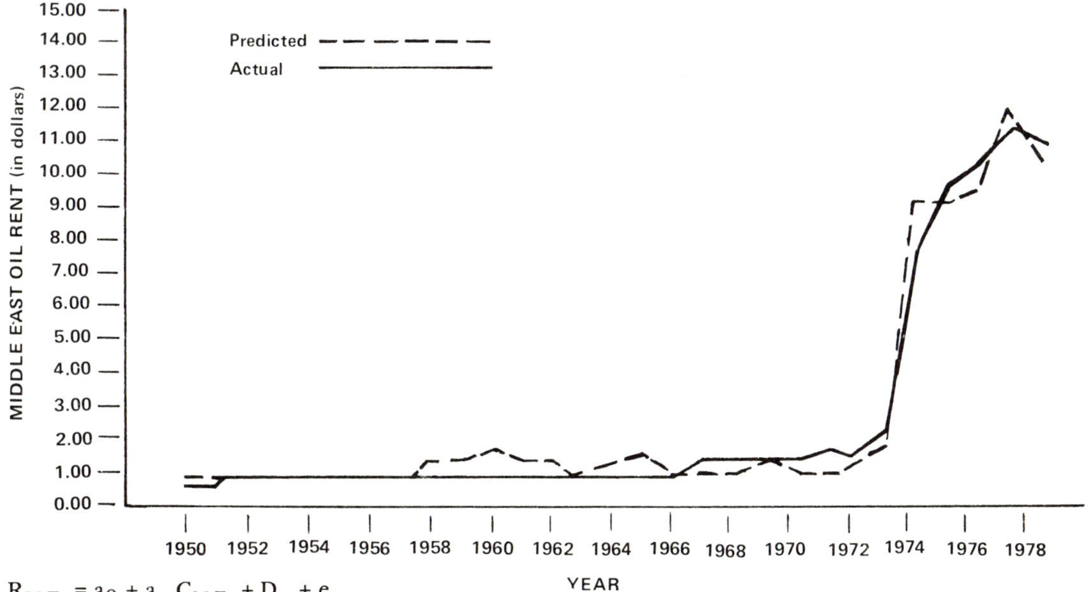

$R_{M.E.} = a_0 + a_1 C_{M.E.} + D_1 + e$,
where $R_{M.E.}$ is oil rent per barrel. $C_{M.E.}$ is the oil capital cost per barrel, and D is the change in the intercept after 1974.

Source: Chase Manhattan Bank, *Capital Expenditures in the World Petroleum Industry* (New York: Chase Manhattan, various issues); OPEC, *Annual Statistical Bulletin*, various issues.

Figure 6

The Actual and Predicted Trends of the Oil Rent per Barrel in the U.S. Oil Region, 1950–1978

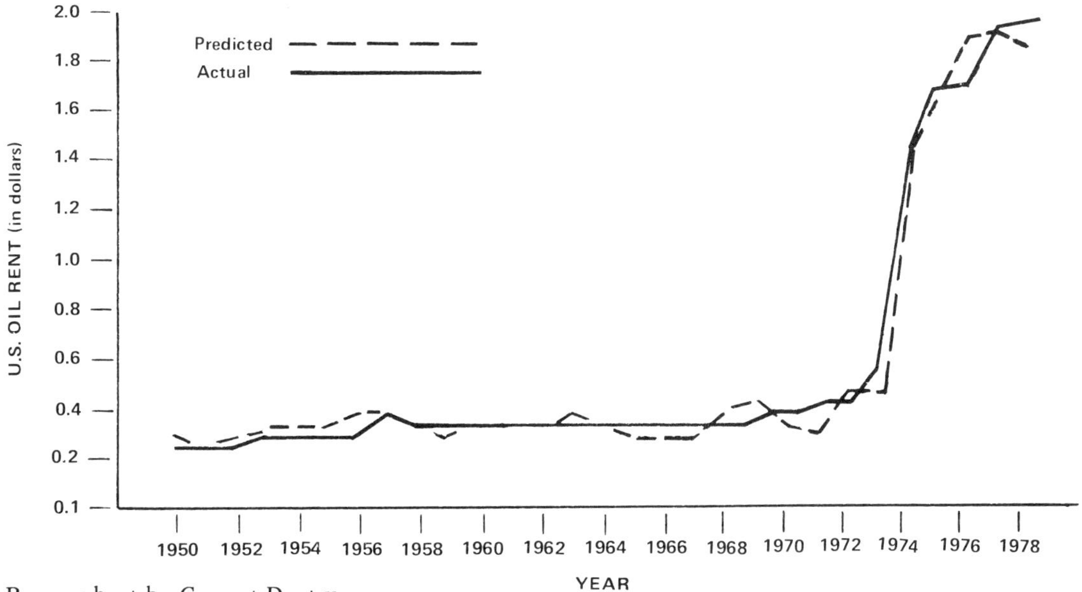

$R_{U.S.} = b_0 + b_1\ C_{U.S.} + D_1 + v,$
where $R_{U.S.}$ is the U.S. rent per barrel, $C_{U.S.}$ is the total U.S. capital cost per barrel, and D_2 is the change in the intercept after 1974.

Source: API, *Basic Petroleum Data Book,* API, various issues.

and the oil prices during the entire period of 1947–75.

The oil rent equations exhibit the characteristics of the most and least productive oil regions of the world. They also illustrate that there was a *'moderate'* level of differential oil rent in the international oil industry prior to the crisis. However, with the crisis, the level of differential oil rent increased. More specifically, the oil rent models show that, following the crisis, the amount of differential oil rents both for the United States and the Middle East increased significantly. Nevertheless, the oil rent increase associated with the Middle East region is by far the largest.

In sum, the price of oil increased in association with the unprecedented increase in the cost of oil recovery from the aged U.S. oil fields. The individual values produced in these fields are the industry value through competition and in conjunction with the internationalization of production within the oil industry at the world level. As a result, individual values formed in the more productive oil regions of the United States, as well as those in all other oil regions of the world, are able to appropriate oil rents (corresponding with their productivities).

The above analysis indicates that the oil crisis did not emerge *because* of the unilateral increase of the 'posted prices' by OPEC. The increase in the 'posted prices' is rather the *consequence* of the structural changes that have taken place in the formation of value in the oil industry. Since the Middle East is among the most productive oil regions of the world, the magnitude of differential oil rents associated with this region is by far the largest in the industry.

NOTES

1. See Appendix A for the detailed computations.
2. J. Johnston, *Econometrics Methods* (New York: McGraw-Hill Book Company, 1972), p. 206. See also Gregory C. Chow, 'Tests of Equality Between Sets of Coefficients in Two Linear Regressions', *Econometrica,* Vol. 28, No. 3, (July, 1960), pp. 591–605.
3. Since the trend of the primary stocks of crude oil for the period of 1947–75 is fairly steady, the figures of excess demand (supply) can be obtained from the direct subtraction of supply and demand quantities. (See Appendix B.)
4. For the above purpose, one has to divide the entire period of 1947–75 into the sub-periods of pre-crisis and the crisis. Since the objective is to calculate the probable changes that might have taken place during the period of crisis, one has to assign a column of zeros to the 1947–71 period, while at the same time one designates the appropriate investment (per barrel) figures for the remaining period, i.e., the period of crisis. This new column of an independent variable in the regression will result in a relationship that, if proved to be significant, will measure the extent of the differential change in the coefficient of the appropriate capital investment per barrel. For further analysis see Jan Kmenta, *Elements of Econometrics* (New York: Macmillan Publishing Co. Inc., 1971), pp. 419–25.
5. Discussing this point with Anwar Shaikh, he has quite correctly pointed out

that the above seems to be technically similar to the time to time adjustments of fixed exchange rates among the national currencies at the international level.

The statistical data used in this chapter are:

The number of exploratory oil wells drilled in the United States (1947–75).
The success rate of exploratory wells drilled in the United States (1947–75).
The number of producing oil wells drilled in the United States (1947–75).
The success rate of total oil wells drilled in the United States (1947–75).
The number of total oil and gas wells drilled in the United States (1947–75).
The total number of United States exploration and development expenditures 1947–75).
The total amount of United States oil discoveries, extensions and revisions 1947–75).
The United States new oil field discoveries (1947–75).
The United States new oil reservoir discoveries in the old fields (1947–75).
The amount of United States oil recovery through revisions and extensions 1947–75).
The United States daily oil refinery input (1947–75).
The United States daily domestic oil production (1947–75).
The United States oil imports (1947–75).
The United States primary stock of crude oil, end-year (1947–75).
The amount of United States oil royalty per barrel (15 per cent of the price) 1947–75).
The Middle East oil capital expenditures (1950–78).
The amount of oil added to the Middle East reserves, 1950–64.
The Middle East oil production, 1950–78.
The amount of indicated gross oil found in the Middle East, 1950–78.
The payment per barrel of oil to the Middle East oil-exporting countries, 1950–78.

The statistical data of the United States crude oil are obtained directly from the United States government and the industry sources.

CHAPTER IX

DEVELOPMENT OF AN ALTERNATIVE THEORY OF OIL CRISIS

1. Introduction

In order to develop an alternative theory of the oil crisis, one needs to start from the critique of the prevailing views on the subject of the oil crisis of 1973-74. As we have seen earlier, every conceivable explanation is involved with describing the actual events and facts corresponding with the period leading to the oil embargo. However, none of these theories has shown the *relation* of these outward appearances to the inner *essence* of the oil crisis. The discovery of inner essence is always through the examination of appearances, but what is lacking in these theories is the proper identification of the *law of motion* of the crisis that has necessarily evolved through a number of observable appearances. It would appear they all missed the underlying cause of the crisis because they took the effects of the crisis at their face value. Thus, the existing theories of the oil crisis, from the most conservative to the most 'radical' ones, suffer from the dominance of *idealism* and the lack of proper perspective.

The antithesis of such theories, therefore, is a necessary framework for building of an alternative theory which has none of the above shortcomings and therefore explains the phenomenon of the oil crisis scientifically. To do so one needs to develop a fundamental theoretical basis for explaining the inner nature of the oil crisis.

In section 2, the basis of the Theory of Oil Rent will be developed. Section 3 will discuss some elements of the alternative theory of the oil crisis. Finally, in the last section the alternative theory will be presented in its entirety.

2. Theory of Oil Rent and Development of Capitalism in the Middle Eastern Oil Industry

Earlier in the text we gave an historical account of the Middle East oil development. We divided the entire period into three distinctive episodes: (1) the early oil concessions, (2) the era of transition, and (3) the period of production and formation of values, oil rents, and market prices on a full-fledged capitalistic basis. The significance of such a division becomes apparent when one periodizes the life span of the entire industry, from the standpoint of the production of surplus value.

At the turn of the present century, pre-capitalist social relations were still dominant in the Middle East. At the same time, movements of international capital into this particular region were increasingly underway. The outcome of the above situation in the oil areas of the Middle East led to the system of early oil concessions. These early concessions contained more or less uniform principles for the surrender of oil property rights to international oil companies or individuals interested in oil exploration.[1] As we have pointed out, the size and duration of such concessions were large and long. The granting parties with whom these contracts were made were the ruling authorities of the pre-capitalist societies of the Middle East. In these societies, the dominant form of property in land was public, which was usually belonging to the State. That is the reason why there are two different sets of oil property laws that would separately govern the oil production in the United States and the Middle East. The size and duration of such contracts themselves are the consequence of the existence of pre-capitalist public property relations in the early development of oil production in the Middle East.

The terms of the concessions, such as the determination of *royalty* and additional payments to the contracting governments and ruling authorities, were also different from what we experience today. Royalty is defined as the value of a portion of oil extracted that goes to the individual owner of the subsoil or the state. This percentage is usually figured at 12.5 per cent. However, at the beginning of oil production, this royalty was denominated to shilling per ton or cent per ton, for all practical purposes. These payments resulted from the surrender of oil property rights to the oil companies. The fixity of payment and the lack of any connection with the price of oil are among the distinguishing features of the concessions of this period. It should be realized that even though there were 'profit sharing' clauses in some concessionary agreements, they often remain unobserved and were never honoured in practice.[2]

The above characteristics of concessions and the prevailing social character of pre-capitalist production in the early years of oil production are the material basis for the development of oil rent in the Middle East. Such a phenomenon, i.e., oil rent, that was associated with the legal form of landed property, is the product of the existing social relations within which oil leases were negotiated and oil exploration and production took place. The separation of mineral property and property in mines, in general, and that of oil property and the oil field property in particular, in the early period resulted in the emergence of historically specific relations of production in the international oil industry. These relations, in turn, have resulted in the production of values in conjunction with the *oil rent.*[3] However, during the period of the early concessions, oil rents were still in existence in their *primitive* and undeveloped form corresponding with the social characteristics peculiar to the early development of capitalism in the Middle Eastern oil industry.

The reflections of these early developments under the *direct domination* of international capital can be observed from the moderate terms of concessions, *static* mechanism of the payment itself, and the *arbitrary* determination of royalties and other associated payments. In essence, the notion of *oil rent* during this era is not yet fully developed to conform to social relations under capitalism in the international oil industry. Therefore, one should not expect to observe a fully articulated structure of property relations that would pose a barrier to accumulation in the oil industry. That is why there was no *economic crisis* during this particular period.

As international capital penetrated deeper into the sphere of oil production in the Middle East, the corresponding social relations for further penetration tended to develop accordingly. The economic and political developments of the post-World War II period, in general, and the development of the Middle Eastern oil industry, in particular, led to conditions that sharpened the contradictions and gradually strengthened the capitalist social relations in the entire region. In this connection, the production of oil increased substantially, and with it the terms under which oil property rights surrendered were accordingly revised. This era started with the establishment and spread of the regime of 50-50 profit sharing in 1949. This newly-devised mechanism has become associated with the notion of 'posted prices' for the actual calculation of oil rents. The 'posted price' was used as a variable basis for the determination of royalty and all other associated payments to the oil-exporting states in the Middle East and elsewhere. At the same time, such a price has been periodically used as a basis of internal transfer of crude oil within the network of major oil companies and their subsidiaries.

Looking through the history of this period, i.e., the period of 1950–1970, one can recognize the heightened level of contradiction which is often associated with capitalist production. The primitive notion of oil rent was transformed into a more developed phenomenon as capitalism developed further, but such a transformation did not occur without the associated political and economic tensions. On the one hand, there were countries that resented the rule of international capital in its entirety, and posed a threat to the hegemony of the international oil companies. The 1951 nationalization of oil in Iran is the reflection of such a contradiction. This antagonistic confrontation—that aimed to put an end to the British domination of the oil industry in Iran—proved that, when it comes to insistence on *national sovereignty*, the name of the game is none other than *gun-boat diplomacy*. Here, the *private* interest of an oil company suddenly becomes the *public* interest of the international capital. On the other hand, there was a great deal of antagonism within the sphere of international capital, which resulted from the increased development of capitalism.

During this period, i.e., the era of transition, social relations of capitalism in the Middle Eastern oil industry were perfected. At the end of this period, value production became the essence of the social process, and in this connection oil rents and market prices developed on a full-fledged capitalistic basis. The extent of the socialization of production at the end of this period was reflected in the formation of value at the level of international oil industry, in conjunction with the global determination and formation of oil rents and prices. In other words, the structure of production and the social basis of its renewal in the entire international oil industry have changed drastically. The above developments were not without contradictions and conflicts between *the owners of oil property* and *the owners of oil fields.* These contradictions have resulted in the transformation of the entire industry and marked the beginning of the development of regulating market values, oil rents, and prices in the international oil industry.

During the early 1970s, due to the high level of socialization in the production process, and internationalization of all circuits of capital within the international oil industry, an organic determination of value, oil rent, and prices developed. This was the first time in the history of oil production that the least productive and the most productive regions confronted one another through competition and in a unique social process of value formation at the global level.

As we have pointed out, the least productive oil producing region is the U.S., as opposed to the Middle Eastern oil region which is the most productive.[4] Coupled with these conditions, the U.S. oil productivity has declined since 1972. Thus, even at the national (i.e., the U.S.) level, the magnitude of regulating value, and with it the differential oil rent and prices, must have increased accordingly. As we have demonstrated earlier, in this period value and with it the price formation is no longer an isolated national matter. Therefore, the decline of the oil productivity in the United States should necessarily result in a higher magnitude of value, differential oil rent, and prices at the international level.

Since the magnitude of regulating market value is now formed on the basis of the least productive oil region of the world (the United States), given the productivity of the Middle East oil region, the latter has been in a position to appropriate the difference as differential oil rent. Such an appropriation is not a characteristic of OPEC alone. Other oil producing governments also tended to exhibit the same type of action given the existing oil property relations within their jurisdictions. Thus, additional implication of the present theoretical exposition is the fact that even if OPEC, as an organization will ever be dissolved, differential oil rents will continue to be appropriated by individual oil countries separately.

3. *An Alternative Theory of Oil Crisis: Antithesis of the Existing Thought*

As we have pointed out, an adequate alternative theory of the oil crisis is the one that emerges through a critical examination of the existing views of the oil crisis of 1973-74. The central theme of such a theory should reflect the internal development of the international oil industry, and the underlying forces that led to the crisis. Another aspect of the theory should be the explanation of the timing of the oil crisis. We have seen that none of the existing theories were able to demonstrate, for instance, why oil prices have changed so drastically, except through the explanations such as monopoly or cartel. But the underlying forces of monopoly or cartel during the time of the crisis were not identified, except through the notion of 'dependence' of the United States on foreign oil. Thus, the static notions of monopoly, cartel, and subjective determination on the part of the U.S. government, were among the reasons given by most theorists of the oil crisis. At this juncture, our intention is to build upon the analysis of the previous section in order to secure an objective understanding of the 1973-74 drastic price change in the international oil industry.

The oil crisis of 1973-74 is the result of the restructuring of production in the international oil industry. As such, it was an *economic crisis* which objectively emerged throughout the entire industry. This crisis was the *first* economic crisis of the international oil industry and was different from previous shortages and supply interruptions which did not change the fundamental social relations of the oil industry. As formulated in the previous section, the entire history of oil production at the international level should be divided into three different periods. The first period corresponds with the primitive character of the early oil production; the second period is the era of transition to a *full-fledged* capitalistic production; and finally, the third period reveals the final consolidation of capitalist production. The law of value is the organizing theme of any capitalist production in general, and that of oil production in particular. More importantly, in the era of imperialism, the importance of the law of value does not diminish but gains further relevance.

Likewise, competition, as an integral part of the operation of the law of value, is relevant to the era of monopoly capitalism. Monopoly and competition should not be distinguished from each other quantitatively, as has so often been done by the majority of the theorists of the oil crisis. Monopoly and competition are part and parcel of a synthesis, and through the accumulation process, one leads to the other. Therefore, it is possible for us to recognize the fact that since *monopoly* existed long before the oil crisis, there is no reason to ascribe a determining role to it. At the same time, our theory of oil rent, developed from the social conditions of accumulation and the historical developments of production at the international level, does reflect the essential characteristic of barrier to accumulation in the oil industry.[5] The separation of property in oil and property of oil fields itself tends to obstruct competition in the oil industry, just as landed property poses a barrier in agricultural production. In this context, the process of the determination of 'posted prices' within the context of value formation at the industry level can be clearly understood. There is no need for the argument of 'economic nationalism' of the OPEC nations; especially 'nationalism' of comprador capitalist states and puppet governments.

The 'posted prices' are the mechanism through which oil rents are determined and appropriated by OPEC. In turn, the magnitude of differential oil rent depends upon the differential productivity of the least and the most productive oil regions of the world. The above process is meaningless without the priority of value production and value formation in competition. Thus, what is wrongly attributed to the discrete and arbitrary decision of OPEC (that is allegedly guided by 'economic nationalism') or 'the OPEC offensive' is really the reflection of the determination of oil rent in relation to the production of value and its reproduction through crisis. The decline in the U.S. oil production is also the reflection of production conditions of the U.S. oil fields. Also the formation of value within the international oil industry necessarily reflects the conditions in the U.S. oil fields while the 'dependence' of the United States on the foreign oil is not the *cause* of the oil crisis, as most theorists tend to believe.

4. *The Alternative Theory of Oil Crisis*

Earlier in the present chapter the attempt has been made to introduce the necessary conceptual, as well as historical elements needed for final presentation of our theory. At this juncture, however, we are prepared to carry out the remaining task.

i. Periodization of Capitalist Development in the Middle Eastern Oil Industry

It has been argued that the history of oil production in the Middle East passed through three different periods. These periods were (1) the era of early concessions (1901–50); (2) the transitional stage of capitalist development (1950–70); and (3) the era of capitalist production and internationalization of capital in the oil industry. The era of early concessions is the period within which international oil companies divided the entire Middle East among themselves by obtaining rights for exploration and production of crude oil, through direct political domination. The areas subjected to these concessions were often close to the size of the countries involved, and such contracts lasted nearly a century. The existence of pre-capitalist social relations along with the lack of private property in land, and the political hegemony of international capital were the dominant features of this period. The second period, however, is associated with the further development of capitalist production in the Middle Eastern oil industry.

Distinguishing features of this period are the usage of 'posted prices', nationalization and subsequent denationalization of the oil industry in Iran, and the formation of the Organization of Petroleum Exporting Countries (OPEC). Meanwhile, along with the development of capitalist social relations and economic dominance of international capital in general, the process of the internationalization of capital in the oil industry accelerated. Finally, during the third period of development, value formation, and in conjunction with it price formation, gained international dimension.

ii. The Nature of Property Relations and the Theory of Oil Rent in the Oil Industry

One of the distinguishing features of oil production relations is the production through the intervention of ownership of property in the oil fields. In other words, the production of oil in the capitalist mode of production is dependent upon the surrender of the rights of ownership of extracted crude oil to capital by the owner of the oil reserves. Thus, the production of oil does historically emerge through the formation of such property relations. The intervention of this barrier in the process of value formation leads to the phenomenon of *oil rent,* just as the *intervention* of landed property in capitalist agriculture leads to agricultural rent.[6]

The phenomenon of *oil rent,* like any other social relation, has passed through different stages of development. During the era of early concessions, when the domination and subsequent expansion of international capital in the Middle East had just started, a *primitive* form of oil property relation was developed in conjunction with the state property in land in most societies of this region. The social basis of oil rent at this stage had not been fully developed to impose a barrier to production. Therefore there was little problem for capital to overcome. In the main the determining condition of the magnitude of oil rents was the direct political hegemony of international capital in the Middle Eastern oil industry during the early period.

The increased development of oil property relations which led to the formation of rent, on the one hand, and growing phenomenon of internationalization of capital, on the other, established an organic relation among the existing oil-producing regions of the world. As the international oil industry extended its sphere of economic and political domination over the entire globe, the extent of this organic relation increased. Meanwhile, oil-production continued under different *production conditions* in each oil producing region of the world. Thus, the magnitudes of *individual values* produced in these regions were necessarily different from each other. As the formation of social value for the entire international oil industry tended to originate from the least productive region of the world, the production in the more productive oil regions received differential oil rent. Figure 7 reveals that the U.S. oil region is the most heavily drilled area of the world. The combination of small oil fields and the high intensity of drilling activity are the significant factors that lead to the high cost of oil development in this region. In contrast, the Middle East oil region shows the highest level of productivity in the international oil industry.

Table 23 illustrates the average daily production per oil well for a selected number of oil-producing countries. An average U.S. oil well produces about 17 barrels per day, whereas the average oil well in Iran produces 11,838 barrels of oil per day. (The average production per well for the entire Middle East is 4,684 daily barrels.) It is evident that the U.S. and the Middle East are located in the opposite ends of the spectrum of productivity in the international oil industry. A comparison of productivity for a selected number of regions (and countries) can be made by examining their cost of finding and developing oil. (Table 24.)

Table 23

Average Daily Production Per Well for a Selected Number of Countries (for late 1960s)

	expressed in daily barrels
United States	17
World	56
Soviet Union	99
Indonesia	248
Venezuela	301
Libya	2,985
Kuwait	4,665
Middle East, all countries	4,684
Iran	11,838

Source: Norman A. White, *Financing the International Petroleum Industry* (London: Graham Trotman, Ltd., 1978), p. 15, (Table 9).

Table 24

Average Cost of Finding and Development of Oil per Barrel for a Selected Number of Regions (1974-1978)

	In U.S. Dollars
United States	4.06
Canada	2.45
Western Europe	1.48
Africa	1.27
Far East	0.90
Western Block	0.88
Venezuela	0.18
Middle East	0.12

Source: *Petroleum Outlook*, May 1979 (vol. 32, no. 5), p. 1.

Figure 7

*Wells Drilled in Relation to Prospective Areas Around the World**

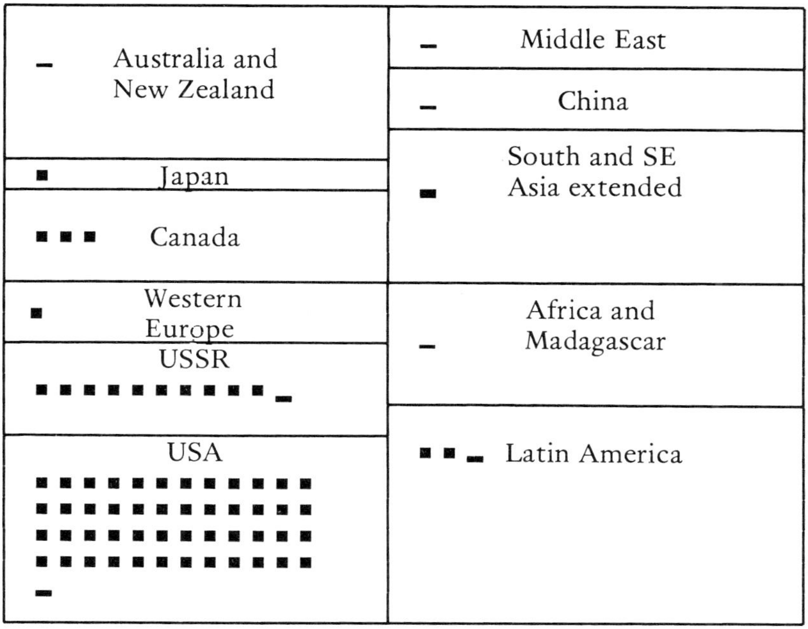

* The areas within the rectangles represent, to scale, the area of petroleum prospects within the country or region named. Both onshore and offshore prospects up to a water depth of 200m are included. The black squares represent the cumulative drilling, exploratory plus development, in each area, up to the end of 1975. A full square represents 50,000 wells and the smaller section proportions thereof. Completion of the remaining area of the United States would correspond to a saturation level of drilling.

Source: R.L. Parker, 'Observation of a Driller', *International Petroleum Times,* April 1, 1979, p. 12.

The average costs of finding and development of oil per barrel in the United States and the Middle East oil region are $4.06 and $0.12 respectively. Again, the U.S. oil region is by far the most expensive area for oil exploration and development.

The theoretical implication of the above investigation is that (1) the formation of the magnitude of value in the international oil industry is based upon the least productive region, (and in this connection, the U.S. oil production is regulating the value, oil rent, and the market prices of crude oil at the international level); and (2) the determination of the payments due to all oil-producing states, including OPEC, is neither arbitrary nor dependent upon the 'behavior of the cartel'. In fact, the task of our theory of oil rent is to explain the significance of the above *distorted appearances* through a more *essential* relation, that is *internal* to oil production, beyond these observable and given forms.

iii. The Process of Value Formation and Price Determination

It should be pointed out that there are many types of prices in the oil industry, and the characteristics of every one of these 'prices' are different from each other. To name a few, there are 'posted prices', 'spot prices', and 'buy-back prices'. From the standpoint of political economy, however, it is necessary to study the nature of 'oil spot prices' and 'posted prices' as well as their relations as to the process of value formation in the oil industry.

The 'spot price' is the price of a single volume of crude oil bought and sold in the international oil market.[7] This price is the one that is usually referred to as the market price in economics. The characteristic of the 'spot price', is its fluctuation around value. Such fluctuations are the result of market conditions, in which demand and supply play an important part.

The 'posted price' is a mechanism used for internal transfer of crude oil within the network of major oil companies and their subsidiaries, and the calculation of oil rent since the 1950s. After the replacement of early oil concessions by the regime of 50–50 profit sharing, the 'posted price' was used as a basis to figure out the share that would go to oil governments. From the historical standpoint, this represented one further step in the development of oil rent. Although the 'posted price' is not a price in an ordinary sense, it has functioned to relate oil rents to the oil prices during the period of transition.

Even though the term 'posted price' has not changed since the time of its adoption in the 1950s, its essential characteristic, as a mechanism in the determination of the oil rents, has considerably changed. These changes correspond with the level of socialization of production, the internationalization of capital, and the development of capitalist social relations in the international oil industry.

One of the significant events during the post-transitional period was the emergence of *spot markets* in crude oil, through which Iranian, Nigerian and Libyan governments were able to receive bids of such magnitude as \$17 to \$24 per barrel during the month of December 1973.[8] As we have observed, the spot price or market price did not decline significantly after the removal of embargo, even though the world economy was moving into a deep recession during the 1974–75 period. Instead, during the embargo period the 'posted price' increased fourfold to the level of \$11.65 for the *marker* crude (Arabian light crude oil of 34° API gravity). The above change increased the magnitude of differential oil rents not only for the OPEC countries, but also for other oil-producing states, such as the United Kingdom. Accordingly, the wellhead price of the so-called 'new oil' in the U.S. increased to the world level.[9]

Since 1970 the international oil industry developed into an organic entity composed of different oil regions of the world. The consequence of such a development was a major reorganization through crisis that swept through the entire structure of the oil industry at the global level. Therefore, the oil crisis of 1973-74 was historically the first *economic crisis* of the oil industry at the international level. Although the above crisis is associated with oil shortages and supply restrictions, one cannot explain it through the phenomenon of oil shortage alone.

As we pointed out earlier, the magnitude of value at the international level for the oil industry is regulated by the condition of production in the U.S., i.e., the least productive producer. At the same time, development capital costs of oil per barrel nearly tripled shortly before the crisis.[10] This increase in costs is the reflection of the increased level of expenditures attributed to the development of older oil fields in the U.S.

In this manner, due to the obstacle of property relations in oil production in general, and the *fragmentation* of new oil leases in particular, the flow of capital investments for the purpose of additional exploration is obstructed. This situation, in turn, diminishes the needed incentives for discovery of oil from the remaining oil lands located in the lower 48 states in the United States.[11] Instead, given the above conditions, investment of capital in conjunction with the previously developed oil fields and old leases are increased. It is precisely due to the increased level of such investments and the resultant decline in their productivity that the cost structures of the aged U.S. oil fields and the magnitude of individual value formed from such fields were increased. As we pointed out, the U.S. oil region has been the least productive area within the international oil industry even prior to the threefold cost increase of 1972 period. The further intensification of capital investment on the existing oil fields of the lower 48 states increased even further the magnitude of capital cost per barrel of oil produced in the U.S. This in turn resulted in the larger value magnitude for the entire international oil industry through the crisis. Since, such a change in the magnitude of value necessarily involved a complete reorganization in the structure of oil production by way of crisis. Therefore, the oil crisis of 1973-74, which was triggered by the October War of 1973, was the condition of the reorganization of the international oil industry and *not its cause.* During this reorganization, the higher value magnitude for the international oil industry (including the U.S.) was formed and in conjunction with it differential oil rents, and market prices increased substantially.

iv. Monopoly and Competition in the Oil Industry

The existence of the dominant but erroneous concept of monopoly is common to all the prevailing views of the oil crisis. Without any exception, all of the existing theories either implicitly or explicitly tend to agree that the price of oil is *directly* determined by the OPEC *cartel*, monopoly oil firms, or both. For instance, the Dependency Theory of the oil crisis implies that the oil prices prior to the events of the 1970s were determined by the oil companies, but after the 'OPEC offensive', which was a move against unequal exchange at the international level, were subsequently set by the oil-exporting countries. The conspiracy theory argues that the increase in the price of oil was initiated by the U.S. government in conjunction with major oil firms and the OPEC countries. The conventional theory of the oil crisis talks about the production, distribution, and marketing controls which do not let the law of supply and demand operate properly. The common denominator of these views is the *quantity theory* of monopoly.[12] The notion of monopoly is perceived to be dependent upon the number of firms within an industry. Accordingly, if the number of firms operating in an industry is small, it is called monopoly or oligopoly. However, if the number of firms in an industry are many, then it is said that probably competition prevails.

In contrast with the above views, since competition and monopoly are necessary in the process of value formation, one cannot talk about monopoly without competition. The formation of value in an industry, in general, and in the oil industry, in particular, emerges through competition. In the oil industry, just like any other industry, competition among different production units, with different *individual values*, leads to the formation of value for the entire industry. However, due to the existence of the barrier of the oil property relation, a portion of surplus value is transformed into differential oil rent and is appropriated by more productive producers. Thus, competition as an inner nature of capital will always be present in the process of accumulation and value formation. The competitive struggle among capitals leads to concentration, centralization and the further monopolization of capital. Competition leads to monopoly and monopoly leads to a further competitive struggle. As a result, it is hardly surprising to see that the dominant theories fail to recognize the true nature of the oil crisis of 1973-74.

5. Concluding Remarks

We have attempted to develop the antithesis of the prevailing views associated with the oil crisis of 1973-74. In order to explain the oil crisis, we have developed a theory of the internationalization of capital in the oil industry, a theory of differential oil rent, and finally a theory of the oil crisis. We maintained that the internationalization of the oil industry has gone through three historical periods: (1) the era of early concessions, (2) the transitional period, and (3) the period of capitalist production and internationalization. We have demonstrated that the phenomenon of oil rent in the early period was associated with the direct political hegemony of international capital, the predominance of pre-capitalist social relations, and the existence of state or non-private property in land. With the development of social relations of capitalist production, oil rents became *value determined.* The theory of differential oil rent was formulated in order to show the extent of the organic relations of different oil regions of the world in relation to each other through competition. We have argued that due to the intervention of the structures of property relations, i.e., the separation of oil property from ownership of the oil reserves, the individual value produced by the least productive producer became the regulating market value of the industry. The more productive producers were in a position to appropriate a portion of surplus value in the form of differential oil rent. Therefore, the regulating value of oil at the industry levels determines the magnitude of differential oil rents and market prices. The 'posted price' is the key mechanism for the determination of differential oil rents. Finally, we have demonstrated that the production in the U.S. oil region became the basis for the formation of the regulating market value for the whole international oil industry. During the period of 1971–72, i.e., prior to the oil crisis, the capital costs of oil increased nearly threefold. As a result, there was a sweeping change in the structure of the entire industry which in turn led to the higher magnitudes for the value, differential oil rents, and market prices at the global level.

Taking the entire energy industry as a whole, it would be appropriate to conclude that the production conditions associated with the exploration and development of fossil fuels tend to *regulate* the market value of other sources of energy. There are three major types of fossil fuels that have substantial shares in the production of energy in the world. These are crude oil, coal, and natural gas. Contrary to the opinion of some of the theorists, we maintain that it is the individual value of crude oil from the aged U.S. oil fields that would regulate the market values of all these major sources of energy globally.

Our argument has two dimensions. The first dimension, which is theoretical, deals with the distinction between organic composition of capital (OCC) and technical composition of capital (TCC) in the U.S. coal and oil industries. It is pointless to confuse these categories by concluding that since '. . . a much greater mass of capital is necessary to produce a given mass of energy' from coal, the individual value of coal must be larger, and therefore this value should necessarily regulate the values of other sources of energy.[13]

The second dimension of our argument relates to *empirical facts* that indicate the price of a particular mass of energy (i.e., BTU) produced both from coal and natural gas at least since the 1950s has been lower than that of oil in the United States.[14] Therefore, on both theoretical and empirical grounds, it is the individual value of crude oil, from the aged U.S. oil fields, that is the basis of the market value of the entire energy industry at the global level. It is in this particular sense that the oil crisis is indeed an *energy crisis.*

NOTES

1. William Knox D'Arcy was the first individual to acquire oil property rights in Persia in 1901.
2. The Anglo-Persian Concession of 1914 and Anglo-Iranian Concession of 1932 both had a clause of profit-sharing, that was frequently a prime source of dispute from the outset and ultimately led to the Iranian nationalization in 1951.
3. We use the notion of rent to reflect the existing social relations. This phenomenon includes 'royalty', and all the associated payments to the host governments by the oil companies.
4. See Tables 23 and 24 in the present chapter and Table 10 in Chapter VII for comparison.
5. Speaking of rent is not new in economics. However, what makes rent an important issue is its specificity in particular types of production. As we have seen in the previous chapters, Marx developed his theory of Agricultural Rent from the analysis of the landed property in agriculture. Therefore, rent in agriculture is the effect of the landed property, and by no means general in political economy. Accordingly, in order to be able to analyze the phenomenal form of rent associated with the oil production, one has to analyze the nature of oil property relations in the oil industry. Consequently, our Theory of Oil Property Relations and Theory of Differential Oil Rent, should be considered as the basis of the Theory of Oil Crisis.
6. Regardless of their apparent similarity, oil production and agricultural production cannot be subjected to one theory of rent, because the notion of rent is the result of the barrier of property relations, which have emerged under different historical and material conditions. Therefore, there is no general theory of rent as such in political economy, and Marx's theory of rent is specific to agriculture. In addition, agricultural rent itself is the *effect* of a specific property relation which cannot be generalized for all others. In fact, the above argument draws a demarcation line between Marx's economic theory and its neo-Ricardians counterpart.
7. Laurence Minard, 'On the Spot', *Forbes,* January 1980, pp. 29–30.
8. Edith Penrose, 'The Development of Crisis', *DAEDALUS,* Fall, 1975, p. 51.
9. Due to the price regulation in the U.S., two categories of crude oil were recognized and a two-tier price system was established. 'old oil' was the volume of crude oil extracted from oil wells that were discovered prior to 1972, while 'new oil' was the oil produced from newly-discovered oil fields or the volume extracted from the old oil wells that was over and above 1972 production level.
10. See Chapter VIII, the results of the econometric models.
11. For an empirical analysis of the fragmentation of oil leases see Edward Miller, 'Some Implication of land Ownership Patterns for Petroleum Policy', *Land Economics,* Vol. 29, (November 1973), pp. 414–423.
12. See John Weeks, *Capital and Exploitation* (New Jersey: Princeton University Press, 1981), pp. 152–58.

13. M. Massarrat, 'The Energy Crisis: The Struggle for the Redistribution of Surplus Profit from Oil', in *Oil and Class Struggle,* edited by Petter Nore and Terisa Turner (London: Zed Press, 1980), p. 34.
14. See Department of Energy, *Annual Report to Congress,* Vol. 2, May 1982, p. 21.

APPENDIX A

LONG-TERM COST CALCULATIONS OF DIFFERENT SPHERES OF U.S. OIL PRODUCTION

In order to examine the relationship between the trend of oil prices, on the one hand, and that of the long-run total capital costs of oil in the United States on the other, one has to calculate the statistical trend of the total capital expenditures per barrel. By the same token, in order to determine the association of the price of oil with the development cost per barrel of the old oil fields, as well as the capital costs per barrel of new oil discoveries, one needs to have the separate cost trend for each sphere of production. The statistical data of the separate capital cost related to the above spheres are not readily available. Therefore, the attempt has been made to calculate these costs from the published reports of the U.S. government and the oil industry.

Table 25 shows the step-by-step calculation of total capital costs, oil exploration costs, and oil development costs per barrel for the period of 1947-75.

Table 25

Calculation of the Total Capital Costs, Exploration Costs, and Development Costs per Barrel of Oil in the United States 1947–1975

(1) Year	(2) Number of Producing Exploratory Oil Wells Drilled (1000)	(3) Success Rate of Exploratory Wells Drilled (per cent)	(4) (2/3x100) Total No. of Oil and Dry Exploratory Wells (1000)	(5) No. of Total Producing Oil Wells Drilled (1000)
1947	0.98	20.3	4.837	17.48
1948	1.10	18.3	6.014	21.76
1949	1.41	20.2	6.980	21.35
1950	1.58	19.5	8.102	23.81
1951	1.76	18.9	9.312	23.18
1952	1.78	18.8	9.468	23.29
1953	1.98	20.1	9.851	25.32
1954	1.98	20.7	9.613	28.14
1955	2.24	20.8	10.769	30.43
1956	2.27	19.1	11.885	30.53
1957	1.94	19.1	10.209	27.36
1958	1.74	19.4	9.021	23.77
1959	1.70	19.8	8.586	24.04
1960	1.32	18.7	7.059	22.26
1961	1.16	17.9	6.480	21.44
1962	1.21	18.4	6.576	21.73
1963	1.31	18.5	7.081	20.14
1964	1.22	16.6	7.305	19.90
1965	0.95	15.4	6.169	18.06
1966	1.20	18.4	6.522	16.78
1967	0.99	17.1	5.789	15.33
1968	0.95	16.2	5.864	14.33
1969	1.08	17.5	6.171	14.37
1970	0.79	16.5	4.787	13.02
1971	0.65	15.7	4.140	11.86
1972	0.68	17.0	4.000	11.31
1973	0.62	20.3	3.054	9.90
1974	0.81	23.3	3.476	12.78
1975	0.97	23.?	4.163	16.41

Source: API, *Basic Petroleum Data Book*, vol. 1, no. 1, Section III, Table 2, Section V, Table 9; API, *Reserves of Crude Oil, Natural Gas Liquids, and Natural Gas in the United States and Canada as of December 31, 1979*, vol. 34 (June, 1980), p. 24, Table II; Department of Energy, *Annual Report to Congress*, vol. 2, various issues.

Table 25 (continued)

Year	(6) Success Rate of Total Wells Drilled (per cent)	(7) (5/6x100) No. of Total Oil and Dry Wells Drilled (1000)	(8) No. of Total Oil and Gas Wells Drilled (1000)	(9) (7/8) Proportion of Wells in Total Oil and Gas (per cent)	(10) (4/7) Proportion of Exploratory in Total Oil Wells
1947	69.0	26.048	30.842	0.844	0.185
1948	69.6	33.262	37.508	0.887	0.181
1949	66.2	32.251	37.656	0.864	0.216
1950	64.8	36.744	42.030	0.874	0.220
1951	61.0	38.000	43.640	0.871	0.245
1952	60.1	38.752	44.560	0.870	0.244
1953	61.4	41.238	47.740	0.864	0.239
1954	63.0	44.667	51.110	0.874	0.215
1955	62.9	48.378	55.150	0.877	0.223
1956	61.3	49.804	57.170	0.871	0.239
1957	61.2	44.706	52.000	0.860	0.228
1958	61.3	38.776	46.940	0.826	0.233
1959	60.9	39.474	47.560	0.830	0.217
1960	60.1	36.988	45.620	0.812	0.191
1961	60.8	35.272	44.250	0.797	0.184
1962	61.3	35.449	44.160	0.803	0.185
1963	59.6	33.792	41.470	0.815	0.209
1964	58.2	34.210	42.290	0.809	0.214
1965	58.2	31.048	38.770	0.801	0.199
1966	58.1	28.881	36.380	0.794	0.226
1967	58.9	26.027	32.230	0.807	0.222
1968	58.1	24.664	30.600	0.806	0.238
1969	57.3	25.078	32.190	0.779	0.246
1970	60.0	21.700	28.120	0.772	0.221
1971	60.7	19.539	25.850	0.756	0.212
1972	59.5	19.008	27.290	0.696	0.210
1973	61.2	16.176	26.590	0.608	0.189
1974	63.2	20.221	31.700	0.638	1.174
1975	64.4	25.481	37.240	0.684	0.163

Table 25 (continued)

Year	(11) Total Oil and Gas Capital Expenditures (Millions) of Dollars	(12) (11x9) Total Oil Capital Costs Millions of Dollars	(13) (12x10) Oil Exploration Costs Millions of Dollars	(14) (12-13) Oil Development Costs Millions of Dollars	(15) Total Oil Discoveries (Millions of Barrels
1947	1950	1645.80	304.473	1341.327	2464.570
1948	2300	2039.64	369.175	1670.465	3795.207
1949	2215	1913.76	413.372	1500.388	3187.845
1950	2325	2031.58	446.948	1584.632	2562.685
1951	2975	2590.63	634.704	1955.926	4413.954
1952	3625	3152.66	769.249	2383.411	2749.288
1953	3975	3433.60	820.630	2512.970	3296.130
1954	4300	3757.77	807.920	2949.850	2873.037
1955	4675	4100.91	914.503	3186.407	2870.724
1956	5075	4421.34	1056.700	3364.640	2974.336
1957	5100	4384.47	999.659	3384.811	2424.800
1958	4225	3490.27	813.233	2677.037	2608.242
1959	4350	3610.50	783.478	2827.022	3666.745
1960	4200	3410.40	651.386	2759.014	2365.328
1961	4000	3188.40	586.666	2601.734	2657.567
1962	4425	3551.95	657.111	2894.839	2180.896
1963	4125	3361.05	702.459	2658.591	2174.110
1964	4450	3599.60	770.314	2829.286	2664.767
1965	4210	3371.37	670.903	2700.467	3048.079
1966	4250	3374.07	762.540	2611.530	2963.978
1967	4365	3524.74	782.492	2742.248	2962.122
1968	5390	4344.34	1033.953	3310.387	2454.635
1969	5250	4090.27	1006.206	3084.064	2120.036
1970	4775	3684.87	814.356	2870.514	3088.918[1]
1971	3900	2947.62	624.895	2322.725	2317.732
1972	6480	4513.32	947.797	3565.523	1557.848
1973	8140	4952.38	936.000	4016.380	2145.831
1974	12355	7881.25	1371.337	6509.913	1993.573
1975	10250	7013.05	1143.127	5869.923	1318.463

[1] The oil reserves discovered in Prudhoe Bay, Alaska is excluded from 1970 oil recovery in order to find a usual picture of reserve addition in the United States. The amount of oil discovery in Prudhoe Bay is estimated at 9.6 million barrels.

Table 25 (continued)

Year	(16) Total New Oil Disco-veries (Millions of Barrels	(17) Total Oil Development (Extensions & Revisions) (Millions of Barrels)	(18) (12:15) Total Oil Capital Costs Per Barrel (Dollars)	(19) (13:16) Oil Explo-ration Costs Per Barrel (Dollars)	(20) (14:17) Oil Develop-ment Costs Per Barrel (Dollars)
1947	445.430	2019.140	0.668	0.684	0.664
1948	396.481	3398.726	0.537	0.930	0.491
1949	890.417	2297.428	0.600	0.464	0.653
1950	564.916	1997.769	0.793	0.791	0.793
1951	389.256	4024.698	0.587	1.630	0.486
1952	496.428	2252.860	1.147	1.549	1.058
1953	591.680	2704.450	1.042	1.387	0.966
1954	585.806	2287.231	1.308	1.379	1.290
1955	476.957	2393.767	1.428	1.917	1.331
1956	467.222	2507.114	1.486	2.262	1.342
1957	416.197	2008.603	1.808	2.402	1.685
1958	314.729	2293.513	1.338	2.584	1.167
1959	369.362	3297.383	0.985	2.126	0.857
1960	253.856	2111.472	1.442	2.566	1.307
1961	361.374	2296.193	1.200	1.623	1.133
1962	380.586	1800.310	1.629	1.731	1.607
1963	349.891	1824.219	1.546	2.008	1.457
1964	346.293	2318.474	1.351	2.220	1.221
1965	471.947	2576.132	1.106	1.421	1.048
1966	310.422	2653.556	1.138	2.456	0.984
1967	344.686	2617.436	1.190	2.270	1.048
1968	357.746	2096.889	1.770	2.886	1.579
1969	247.184	1872.852	1.929	4.072	1.646
1970	368.637[1]	2720.281	1.193	2.209	1.055
1971	156.710	2161.022	1.272	3.987	1.075
1972	278.430	1279.418	2.897	3.404	2.787
1973	203.913	1941.918	2.308	4.590	2.068
1974	313.726	1679.847	3.953	4.321	3.885
1975	301.064	1017.399	5.319	3.797	5.769

1 See the footnote associated with column 15 of the present table.

APPENDIX B

CALCULATION OF DAILY EXCESS DEMAND FOR CRUDE OIL IN THE UNITED STATES

In order to estimate the trend of excess demand (supply) of oil for the period of 1947–75, one needs to have the market demand and supply of the crude oil in United States. The crude oil demand is the amount of oil refinery input (crude oil runs to stills). However, the supply of crude oil is the composition of domestic oil production and oil imports. Meanwhile, since the quantity of domestic supply can be increased or decreased with the change in the quantity of primary stocks (inventories) of crude oil, one has to accommodate for probable changes that might have taken place in the latter quantity. However, data for primary stocks of crude oil exhibit a constant trend for the period under study. Therefore, the actual trend of domestic supply can be estimated without the consideration of the figures for primary stocks.

Table 26 shows the calculation of daily excess demand for crude oil from the available statistics.

Table 26

Calculation of Daily Excess Demand for Crude Oil in the United States
1947–1975

(1) Year	(2) Average Daily Oil Refinery Input*	(3) Average Daily Domestic Oil Production	(4) Average Daily Oil Imports	(5) Total Oil Supply (3+4)	(6) Primary Stocks of Crude Oil, Year-End	(7) Daily Excess Demand for Crude Oil (2–5)
	Figures are in Millions of Barrels					
1947	5.075	5.087	0.267	5.354	N.A.	-0.279
1948	5.596	5.520	0.353	5.873	N.A.	-0.276
1949	5.237	5.046	0.421	5.467	N.A.	-0.230
1950	5.739	5.407	0.487	5.894	N.A.	-0.155
1951	6.494	6.158	0.490	6.648	256	-0.154
1952	6.670	6.256	0.573	6.829	272	-0.159
1953	7.000	6.458	0.648	7.106	274	-0.106
1954	6.958	6.342	0.656	6.998	258	-0.040
1955	7.480	6.807	0.782	7.589	266	-0.109
1956	7.937	7.151	0.934	8.085	266	-0.148
1957	7.919	7.170	1.022	8.192	282	-0.273
1958	7.606	6.710	0.953	7.663	263	-0.057
1959	7.994	7.053	0.966	8.019	257	-0.025
1960	8.067	7.035	1.015	8.050	240	+0.017
1961	8.184	7.183	1.045	8.228	245	-0.044
1962	8.410	7.332	1.126	8.458	252	-0.048
1963	8.687	7.542	1.131	8.673	237	+0.014
1964	8.807	7.614	1.198	8.812	230	-0.005
1965	9.043	7.804	1.238	9.042	220	+0.001
1966	9.444	8.295	1.225	9.520	238	-0.076
1967	9.815	8.811	1.128	9.939	249	-0.124

Table 26 (continued)

Figures are in Millions of Barrels

(1) Year	(2) Average Daily Oil Refinery Input*	(3) Average Daily Domestic Oil Production	(4) Average Daily Oil Imports	(5) Total Oil Supply (3+4)	(6) Primary Stocks of Crude Oil, Year-End	(7) Daily Excess Demand for Crude Oil (2–5)
1968	10.312	8.660	1.290	9.950	272	+0.362
1969	10.630	8.778	1.409	10.187	265	+0.443
1970	10.870	9.180	1.324	10.504	276	+0.366
1971	11.200	9.032	1.681	10.713	260	+0.486
1972	11.700	8.998	2.216	11.214	246	+0.482
1973	12.430	8.784	3.244	12.028	242	+0.403
1974	12.130	8.375	3.477	11.852	265	+0.281
1975	12.440	8.007	4.105	12.112	271	+0.330

* Data for the period of 1947–69 are obtained from API, *Petroleum Facts & Figures*, 1971 Edition. For the remaining period DOE, *Annual Report to Congress*, vol. 2, 1982, p. 63, is used.

Source: API, *Petroleum Facts & Figures*, 1971 Edition; Department of Energy, *Annual Report to Congress*, vol. 2, 1982, p. 63. Table 28, p. 71, Table 32, API, *Basic Petroleum Data Book*, July, 1976, Section VII, Tables 3 and 3a.

APPENDIX C

ON THE PROBLEMS OF AUTOCORRELATION AND MULTICOLLINEARITY

The following models are tested for autocorrelation:

Equation I $\hat{P} = 1.59 + \underset{(0.46)}{1.21C^d} - \underset{(0.18)}{0.09C^e} + \underset{(0.34)}{0.58DC^d}$

Equation II: $\hat{P} = 1.03 + \underset{(0.44)}{1.39C^t} - \underset{(0.80)}{1.34X} + \underset{(0.31)}{0.60DC^t}$

Equation III: $\hat{R}_{M.E.} = 0.57 + \underset{(1.45)}{9.10C_{m.e.}} + \underset{(0.30)}{8.33D_2}$

Equation IV: $\hat{R}_{U.S.} = 0.31 + \underset{(0.01)}{0.09C_{u.s.}} + \underset{(0.07)}{0.92D_1}$

with $d_I = 2.0057$, $d_{II} = 1.8213$, $d_{III} = 2.3053$, and $d_{IV} = 2.7772$, respectively.

It turns out that equations I, II and III exhibit no significant degree of autocorrelation. In other words, both d_I, and d_{II}, and d_{III} are within the acceptable region at 2.5% significance for each tail. Since the degree of autocorrelation for equation IV is significant, an attempt was made to remove the problem by estimating new parameters for this equation. Therefore, equation VI in the text is our alternative equation. For all the above calculations, the SAS and SPSS statistical packages are used.

In order to find out whether or not there is a multicollinearity problem in the estimation of the coefficients of the above equations, the instrumental variable technique has been used to 'separate the effects that the various regressors have on the independent variable'.[1] It turned out that neither the size nor the degree of significance of coefficients was different from our estimates in the above equations.[2]

1 H.H. Kelejian and W.E. Oates, *Introduction to Econometrics* (New York: Harper & Row, 1974), p. 130.

2 For further analysis see Kelejian and Oates, pp. 128–42.

APPENDIX D

THE MIDDLE EAST TOTAL CAPITAL EXPENDITURES AND OIL INCOME PER BARREL

In order to calculate the Middle East total capital costs per barrel for the period under study, one has to have the figures for the total oil found and the total exploration and development expenditures. The total oil found in the Middle East is the summation of the oil reserves added, plus the quantity of oil produced during the particular years. (See Table 27.)

The figures for indicated gross oil found are calculated for the period of 1950–64. However, for the remaining part of the period under study, the above data are directly obtained from *Petroleum Outlook*, May, 1976, and May, 1979. (See Table 27, Column 4.)

The figures for Middle East oil receipts per barrel are the 'royalty' plus all other associated proceeds appropriated by the oil-exporting governments of the Middle East.

Table 27

The Middle East Total Capital Expenditures and Oil Income per Barrel
1950–1978

(1) Year	(2) Middle East Oil Reserves Added During Year (Millions of Barrels)	(3) Middle East Oil Production (Millions of Barrels)	(4) Indicated Gross Oil Found (Millions of (Barrels)	(5) Total Middle East Oil Capital Costs (Millions of Dollars)	(6) Total Middle East Capital Costs Per Barrel (Dollars)	(7) Middle East Oil Receipts Per Barrel (Dollars)
1950	9154	641	9795	80	0.0082	0.370
1951	9753	701	10454	75	0.0072	0.410
1952	13505	761	14266	130	0.0091	0.530
1953	13335	885	14220	100	0.0070	0.660
1954	19299	999	20298	110	0.0054	0.720
1955	28812	1186	29998	140	0.0047	0.760
1956	18199	1261	19460	205	0.0105	0.780
1957	25096	1293	26389	225	0.0085	0.800
1958	4385	1558	5943	320	0.0538	0.790
1959	7485	1685	9170	325	0.0354	0.790
1960	1724	1923	3647	290	0.0795	0.752
1961	5044	2056	7100	305	0.0430	0.735
1962	5771	2258	8029	300	0.0374	0.722
1963	13393	2486	15879	180	0.0113	0.790
1964	4812	2786	7598	205	0.0300	0.762
1965	(1)	(1)	6234	440	0.0710	0.785
1966	(1)	(1)	23663	305	0.0130	0.800
1967	(1)	(1)	17261	340	0.0200	0.850
1968	(1)	(1)	25687	350	0.0140	0.882
1969	(1)	(1)	6371	380	0.0600	0.860
1970	(1)	(1)	29587	330	0.0110	0.895

Table 27 (continued)

(1) *Year*	*(2)* *Middle East Oil Reserves Added During Year (Millions) of Barrels)*	*(3)* *Middle East Oil Production (Millions) of Barrels)*	*(4)* *Indicated Gross Oil Found (Millions of (Barrels)*	*(5)* *Total Middle East Oil Capital Costs (Millions of Dollars)*	*(6)* *Total Middle East Capital Costs Per Barrel (Dollars)*	*(7)* *Middle East Oil Receipts Per Barrel (Dollars)*
1971	(1)	(1)	19535	525	0.0270	1.205
1972	(1)	(1)	14384	560	0.0390	1.302
1973	(1)	(1)	10100	905	0.0900	2.000
1974	(1)	(1)	18450	1050	0.0570	7.825
1975	(1)	(1)	22600	1075	0.0470	9.820
1976	(1)	(1)	22080	2125	0.0960	10.610
1977	(1)	(1)	6463	2500	0.3870	12.155
1978	(1)	(1)	11418	2750	0.2410	11.595

(1) Since the figures for indicated gross oil found for 1965–78 are directly obtained from *Petroleum Outlook*, there is no step-by-step calculation for the above period.

Sources: API, *Basic Petroleum Data Book*, Section II, Tables 1 and 1a; Section IV, Tables 1 and 1a; *Petroleum Outlook*, May, 1976 and 1979; Frank R. Wyant, *The United States, OPEC and Multinatural Oil* (Lexington, Mass.: D.C. Heath & Company, 1979), p. 169, Tables A-2; C. Issawi and M. Yeganeh, *The Economics of Middle Eastern Oil* (New York: Praeger Publishers, 1962), p. 114, Table 34; Bryan Cooper, *OPEC: Oil Report*, ed. (London: Petroleum Economist, 1979).

BIBLIOGRAPHY

Books and Monographs

Adams, W.G., *The Structure of American Industry.* 4th ed. New York: Macmillan Publishing Company, 1971.

Adelman, M.A., 'The World Oil Outlook', in *Natural Resources and International Development.* Edited by Marion Clawson. Baltimore: Johns Hopkins University Press, 1964.

Adelman, M.A., 'Trends in the Cost of Finding and Developing Crude Oil and Gas in the United States', in *Essays in Petroleum Economics.* Edited by S.L. Gardner and S.H. Hanke. Colorado Springs: Colorado School of Mines Press, 1967.

Adelman, M.A., *The World Petroleum Outlook.* Baltimore: John Hopkins University Press, 1972.

Allvine, F.C. and Patterson, J.M., *Highway Robbery: An Analysis of Gasoline Crisis.* Bloomington: Indiana University Press, 1974.

American Petroleum Institute. *Basic Petroleum Data Book.* Washington, D.C.: API, 1978, 1979, 1980.

American Petroleum Institute. *Petroleum Facts & Figures.* Washington, D.C.: API, 1971.

American Petroleum Institute. *Reserves of Crude Oil, Natural Gas Liquids, and Natural Gas in the United States and Canada as of December 1979.* Vol. 32 (June 1980).

Balestra, Pietro, *The Demand for Natural Gas in the U.S.* Amsterdam: North Holland Publishing, 1967.

Baran, P.A. and Sweezy, P.M., *Monopoly Capital.* New York: Monthly Review Press, 1966.

Bedour, C.R. and Dunnington, H.V., *The Petroleum Geology and Resources of the Middle East.* Beaconsfield, England: Scientific Press Ltd., 1975.

Bettleheim, Charles, 'Theoretical Comments', in Emmanuel, A., *Unequal Exchange.* New York: Monthly Review Press, 1972.

Blair, John M., *Economic Concentration.* New York: Harcourt Brace Jovanovich, 1972.

Blair, John M., *The Control of Oil.* New York: Pantheon Books, 1976.

Bosson, Rex and Varon, Bension, *The Mining Industry and the Developing Countries.* New York: Oxford University Press, 1977.

Bradley, Paul G., *The Economics of Crude Petroleum Production.* Amsterdam: North Holland Publishing, 1967.

The Brookings Institution, *Cooperative Approaches to the Energy Problem.* Washington, D.C.: Brookings Institution, 1974.

Brown, Keith C., *The Regulation of the Natural Gas Producing Industry.* ed. Baltimore: Johns Hopkins University Press, 1972.

Campbell, Robert W., *The Economics of Soviet Oil and Gas.* Baltimore: Johns Hopkins University Press, 1968.

Cassady, Ralph Jr., *Price Making and Price Behavior in the Petroleum Industry.* Petroleum Monograph Series 1. New Haven, Conn.: Yale University Press, 1954.

Cattan, Henry, *The Law of Oil Concessions in the Middle East and North Africa.* Dobbs Ferry, New York: Oceana Publications, Inc., 1967.

Cattan, Henry, *The Evolution of Oil Concessions in the Middle East and North Africa.* Dobbs Ferry, New York: Oceana Publications, Inc., 1967.

Caudill, William W., *A Bucket of Oil: The Humanistic Approach to Building Design for Energy Conservation.* Boston: Cahners, 1974.

Chamberlin, E.H., *The Theory of Monopolistic Competition.* Cambridge, Mass.: Harvard University Press, 1950.

Chase Manhattan Bank, *Capital Investment of the World Petroleum Industry.* New York: Petroleum Department, various years.

Chase Manhattan Bank, *Financial Analysis of a Group of Petroleum Companies.* New York: Chase Manhattan Bank, 1978.

Chevalier, Jean Marie, 'Theoretical Elements for An Introduction to Petroleum Economics', in Market, *Corporate Behavior and the State.* Edited by A.P. Jacquemin and H.W. de Jong. Hague: Martinus Nijhoff, 1976.

Clark, Wilson, *Energy for Survival.* Garden City, New York: Anchor Books, 1974.

Cooper, Bryan, *OPEC: Oil Report.* ed. London: Petroleum Economist, 1979.

Dasgupta, P.S. and Heal, G.M., *Economic Theory and Exhaustible Resources.* Cambridge: Cambridge University Press, 1979.

Davenport, E.H. and Cooke, S.R., *The Oil Trusts and Anglo-American Relations.* London: Macmillan, 1923.

De Chazeau, M.G. and Kahn, A.E., *Integration and Competition in the Petroleum Industry.* New Haven, Conn.: Yale University Press, 1959.

DeNovo, John, *American Interests in the Middle East, 1900–1939.* Minneapolis: n.p., 1963.

Dewey, Donald, *The Theory of Imperfect Competition: A Radical Reconstruction.* New York: Columbia University Press, 1969.

Doscher, Todd M., 'Domestic Oil and Gas Supplies From Improved Recovery', in *Energy Technology,* vol. 5. Edited by Richard F. Hill. Proceedings of the Fifth Energy Technology Conference Febuary 27–

March 1, 1978, Washington, D.C.: Government Institutes, Inc., 1978.

Dunlop, John C., 'Petroleum Exploration in Foreign Countries', in *Economics of Petroleum Exploration, Development, and Property Evaluation.* Prentice-Hall for Southwestern Legal Foundation, 1961.

Edens, David G., *Oil and Development in the Middle East.* New York: Praeger Publishers, 1979.

Elwell-Sutton, L.P., *Persian Oil: A Study in Power Politics.* London: Lawrence & Wishart, 1955.

Emmanuel, A., *Unequal Exchange.* New York: Monthly Review Press, 1972.

Engler, Robert, *The Politics of Oil.* Chicago: University of Chicago Press, 1961.

Enos, J.L., *Petroleum Progress and Profits: A History of Process Innovation.* Cambridge, Mass.: MIT Press, 1962.

Fanning, L.M., *Foreign Oil and the Free World.* New York: McGraw-Hill, 1954.

Federal Energy Administration, *The Arab Embargo Two Years Later.* FEA, 1975.

Feis, Herbert, *Petroleum and American Foreign Policy.* Stanford, Calif.: Food Research Institute, 1944.

Fine, Ben and Harris, Laurence, *Rereading Capital.* New York: Columbia University Press, 1979.

Fine, Ben, *Economic Theory and Ideology.* London: Edward Arnold, 1980.

First National City Bank of New York, *Petroleum in the Eastern Hemisphere.* New York: Petroleum Department, 1959.

Ford Foundation. Energy Policy Committee, *A Time to Choose: America's Energy Future.* Cambridge, Mass.: Ballinger Publishing Company, 1974.

Foster Associates, *Energy Prices, 1960-73.* Cambridge, Mass.: Ballinger Publishing Company, 1974.

Frank, Andre Gunder, *Crisis: In the World Economy.* New York: Holmes & Meier Publishers, Inc., 1980.

Frank, H.J., *Crude Oil Prices in the Middle East.* New York: Praeger Publishers, 1966.

Frankel, Paul H., *The Essentials of Petroleum.* London: Frank Cass, 1961.

Frankel, Paul H., *Mattei, Oil and Power Politics.* London: Faber & Faber, 1966.

Freeman, S.D., *Energy, the New Era.* New York: Vintage Books, 1974.

Fried, E.R. and Schultze, C.L., *Higher Oil Prices and the World Economy: The Adjustment Problem.* Washington, D.C.: The Brookings Institution, 1975.

Gardner, Richard N., *World Food and Energy Crisis.* Rensselaerville, New York: Institute on Man and Science, 1974.

Girvan, Norman, *Corporate Imperialism: Conflict and Expropriation.* New York: Monthly Review Press, 1976.

Hamilton, Daniel, *Competition in Oil.* Cambridge, Mass.: Harvard University Press, 1958.

Hanke, S.H. and Gardner, S.L., *Essays in Petroleum Economics.* Colorado Springs: The Colorado School of Mines Press, 1967.

Hartshorn, J.E., *Oil Companies and Governments.* 2nd ed. London: Faber and Faber, 1967.

Hawkins, Clark A., *The Field Price Regulation of Natural Gas.* Tallahasse, Fla.: Florida State University Press, 1969.

Hewins, Ralph, *Mr. Five Per Cent.* London: Hutchinson, 1957.

Hidy, R.W. and Muriel E., *Pioneering in Big Business: History of Standard Oil Co. (New Jersey), 1882-1911.* New York: Harper, 1956.

Hirst, David, *Oil and Public Opinion in the Middle East.* New York: Praeger Publishers, 1966.

Hobson, C.K., *Export of Capital.* London: Constable, 1914.

Hubbert, M.K., *Energy Resources.* Washington, D.C.: National Academy of Sciences and National Research Council, 1962.

Hurewitz, J.C., *Middle East Dilemmas.* New York: Harper, 1953.

Issawi, Charles P. and Yeganeh, Mohammed, *Economics of Middle Eastern Oil.* New York: Praeger Publishers, 1962.

Issawi, Charles P., *Oil, The Middle East and The World.* Washington, D.C.: Center for Strategic and International Studies of Georgetown University, 1972.

Ise, John, *United States Oil Policy.* New Haven, Conn.: Yale University Press, 1926.

Jacoby, Neil H., *Multinational Oil: A Study of Industrial Dynamics.* New York: Macmillan Publishing Company, Inc., 1974.

Johnston, J., *Econometrics Methods.* 2nd ed. New York: McGraw-Hill Book Company, 1972.

Kelejian, H.H. and Oates, W.E., *Introduction to Econometrics: Principles and Applications.* New York: Harper & Row Publishers, 1974.

Keynes, John M., *The General Theory of Employment, Interest, and Money.* New York: Harcourt, Brace & World, Inc., 1964.

Kim, Young Y. and Thompson, Russell G., *New Oil and Gas Supplies in the Lower 48 States.* Houston, Texas: Gulf Publishing Company, 1978.

Kmenta, Jan, *Elements of Econometrics.* New York: Macmillan Publishing Company, Inc., 1971.

Krueger, Robert B., *The United States and International Oil.* New York: Praeger Publishers, 1975.

Landsberg, Hans H. and Schurr, Sam H., *Energy in the United States.* New York: Random House, 1968.

Leeman, Wayne, *The Price of Middle East Oil.* Ithaca, New York: Cornell

PK 273 LEE

University Press, 1962.

Lenczowski, George, *Oil and the State in the Middle East.* Ithaca, New York: Cornell University Press, 1960.

Lenin, V.I., *Imperialism, The Highest Stage of Capitalism.* New York: International Publishers, 1970.

Lewis, R.S. and Spinrad, B.I., *The Energy Crisis.* ed. Chicago: Bulletin of the Atomic Scientists, 1972.

Longrigg, Stephen H., *Oil in the Middle East.* New York: Oxford University Press, 1961.

Lovejoy, W.F., Homan, P.T., and Galvin, C.O., *Cost Analysis in the Petroleum Industry.* Dallas, Texas: Southern Methodist University Press, 1964.

Lovejoy, W.F. and Homan, P.T., *Problems of Cost Analysis in the Petroleum Industry.* Dallas, Texas: Southern Methodist University Press, 1964.

Lubell, Harold, *Middle East Oil Crisis and Western Europe's Energy Supplies.* Baltimore: Johns Hopkins University Press, 1963.

MacKay, D.I. and MacKay, G.A., *The Political Economy of North Sea Oil.* Boulder, Colorado: Westview Press, 1975.

McDonald, Stephen L., *Petroleum Conservation in the United States: An Economic Analysis.* Baltimore: Johns Hopkins University Press, 1971.

McLean, J.G. and Haigh, R.W., *The Growth of Integrated Oil Companies.* Cambridge, Mass.: Harvard University Press, 1954.

Manoharan, S., *The Oil Crisis: End of An Era.* New Delhi: S. Chand & Co. Ltd., 1974.

Marshall, Alfred, *Principles of Economics.* 8th ed. New York: Macmillan & Co. Ltd., 1961.

Marx, K. and Engels, F., *Selected Correspondence.* Moscow: Progress Publishers, 1965.

Marx, Karl, *Capital,* vol. 3. New York: International Publishers, 1967.

Marx, Karl, *Theories of Surplus Value,* Part I. Moscow: Progress Publishers, 1969.

Marx, Karl, *Theories of Surplus Value,* Part II. Moscow: Progress Publishers, 1968.

Marx, Karl, *Theories of Surplus Value,* Part III. Moscow: Progress Publishers, 1971.

Marx, Karl, *Poverty of Philosophy.* Moscow: Progress Publishers, 1973.

Marx, Karl, *Grundrisse: Foundations of the Critique of Political Economy.* New York: Vintage Books, 1973.

Marx, Karl, *Capital: Critique of Political Economy,* vol. 1. New York: Vintage Books, 1977.

Marx, Karl, *Capital: Critique of Political Economy,* vol. 2. New York: Vintage Books, 1981.

Marx, Karl, *Capital: Critique of Political Economy,* vol. 3. New York: Vintage Books, 1981.

Mead, W.J. and Utton, A.E., *U.S. Energy Policy: Errors of the Past, Proposals for Future.* Cambridge, Mass.: Ballinger Publishing Company, 1979.

Medvin, Norman, *The Energy Cartel.* New York: Vintage Books, 1974.

Mendershausen, H., *Dollar Shortage and Oil Surplus in 1949-50.* Essays in International Finance, no. 11. Princeton, N.J.: International Finance Section, Department of Economics, Princeton University, 1950.

Mikdashi, Zuhayr, *A Financial Analysis of Middle Eastern Oil Concessions: 1901-1965.* New York: Praeger Publishers, 1966.

Mikdashi, Zuhayr, *The Community of Oil Exporting Countries.* Ithaca, New York: Cornell University Press, 1972.

Mikesell, Raymond F. and Chenery, Hollis B., *Arabian Oil: America's Stake in the Middle East.* Chapel Hill, N.C.: The University of North Carolina Press, 1949.

Mikesell, Raymond F., *Foreign Investment in The Petroleum and Mineral Industries.* Baltimore: Johns Hopkins University Press, 1971.

Miller, Roger L., *The Economics of Energy.* New York: Morrow, 1974.

Mosely, Leonard, *Power Play.* New York: Random House, 1973.

Nash, Gerald D., *United States Oil Policy, 1890-1964.* Pittsburgh: University of Pittsburgh Press, 1968.

National Academy of Sciences, *Resources and Man.* National Research Council, 1969.

National Academy of Sciences, *Mineral Resources and the Environment.* National Research Council, 1975.

Neumann, John Von and Morgenstern, Oskar, *Theory of Games and Economic Behavior.* Princeton, New Jersey: Princeton University Press, 1944.

Nore, Petter and Turner, Terisa, ed. *Oil and Class Struggle.* London: Zed Press, 1980.

O'Conner, Harvey, *The Empire of Oil.* New York: Monthly Review Press, 1955.

O'Conner, Harvey, *World Crisis in Oil.* New York: Monthly Review Press, 1962.

O'Conner, Richard, *The Oil Barons.* Boston: Little Brown, 1971.

Odell, Peter, *Oil and World Power, A Geographical Interpretation.* Harmondsworth, England: Penguin, 1970.

OPEC, *Annual Statistical Bulletin.* n.p., various issues.

Organization for Economic Cooperation and Development (OECD), *Energy Policy, Problems and Objectives.* Paris: n.p., 1966.

Organization for Economic Cooperation and Development (OECD), *Energy Prospects to 1985*, 2 vols. Paris: n.p., 1974.

Penrose, Edith, *The Large International Firm in Developing Countries:*

The International Petroleum Industry. London: Allen & Unwin, 1968.

Rybczynski, T.M., *The Economics of the Oil Crisis.* ed. New York: Holmes & Meier Publishers, Inc., 1976.

Radice, H., *International Firms and Modern Imperialism.* ed. London: Penguin, 1975.

Ricardo, David, *The Principles of Political Economy and Taxation.* 3rd ed. London: J.M. Dent & Sons Ltd., 1976.

Richman, P.B., *Taxation of Foreign Investment Income.* Baltimore: Johns Hopkins University Press, 1963.

Ridgeway, James, *The Last Play: The Struggle to Monopolize the World's Energy Resources.* New York: Dutton, 1973.

Rifai, Taki, *The Pricing of Crude Oil.* New York: Praeger Publishers, 1974.

Robinson, Joan, *The Economics of Imperfect Competition.* London: Macmillan, 1933.

Robinson, Joan, *Economic Philosophy.* Garden City, New York: Anchor Books, 1964.

Rocks, Lawrence and Runyon, Richard P., *The Energy Crisis.* New York: Crown, 1972.

Rouhani, Fuad, *A History of OPEC.* New York: Praeger Publishers, 1971.

Ruttenberg, S.H. and Associates, *The American Oil Industry: A Failure of Antitrust Policy.* n.p., 1974.

Safer, Arnold E., *International Oil Policy.* Lexington, Mass.: Lexington Books, 1979.

Sampson, Anthony, *The Seven Sisters.* New York: Bantam Books, 1975.

Samuelson, P.A., *Economics.* 9th ed. New York: McGraw-Hill Book Company, 1973.

Schaffer, Edward A., *Oil Import Program of the United States.* New York: Praeger Publishers, 1968.

Schurr, Sam and Netschert, Bruce, *Energy in the American Economy, 1850-1975.* Baltimore: Johns Hopkins University Press, 1960.

Schurr, S. and Homan, P.T., *Middle Eastern Oil and the Modern World.* New York: American Elsevier, 1971.

Schwadran, Benjamin, *Middle East Oil and the Great Powers.* New York: Praeger Publishers, 1955.

Shaikh, Anwar, 'A History of Crisis Theories', in *U.S. Capitalism in Crisis.* New York: Union for Radical Political Economics, 1977.

Shaikh, Anwar, 'The Poverty of Algebra', in *The Value Controversy.* London: New Left Books, 1981, pp. 266–300.

Shamsedin, Ezzedin M., *Arab Oil and the United States: An Admixture of Politics and Economics.* Columbia, S.C.: The University of South Carolina Press, 1974.

Stigler, G.J., *The Theory of Competitive Price.* New York: The Macmillan Company, 1946.

Stocking, George W., *Middle East Oil.* Nashville: Vanderbilt University

Press, 1970.

Stork, Joe, *Middle East Oil and the Energy Crisis.* New York: Monthly Review Press, 1975.

Tanzer, Michael, *The Political Economy of International Oil and the Underdeveloped Countries.* Boston: Beacon Press, 1969.

Tanzer, Michael, *The Energy Crisis: World Struggle for Power and Wealth.* New York: Monthly Review Press, 1974.

Tanzer, Michael, *The Race for Resources: Continuing Struggles Over Minerals and Fuels.* New York: Monthly Review Press, 1980.

Tugendhat, Christopher, *Oil, the Biggest Business.* New York: Putnam, 1968.

Vallenilla, Luis, *Oil: The Making of a New Economic Order.* New York: McGraw-Hill Book Company, 1975.

Vicker, Ray, *Kingdom of Oil, The Middle East.* New York: Scribner's, 1974.

Weeks, John, *Capital and Exploitation.* Princeton, New Jersey: Princeton University Press, 1981.

White, Gerald T., *Formative Years in the Far West: A History of Standard Oil Company of California and Predecessors Through 1919.* New York: Appleton-Century Crofts, 1962.

White, Norman A., *Financing the International Petroleum Industry.* London: Graham Trotman Ltd., 1978.

Williamson, H.F. and Dawn, A.E., *The American Petroleum Industry: The Age of Illumination, 1855-99.* Evanston, Ill.: Northwestern University Press, 1959.

Williamson, H.F., Andreano, R.L., Daum, A.E., and Klose, G.C., *The American Petroleum Industry: The Age of Energy, 1899-1959.* Evanston, Ill.: Northwestern University Press, 1963.

Wyant, Frank R., *The United States, OPEC, and Multinationals.* Lexington, Mass.: Lexington Books, 1977.

Zimmerman, E.W., *Conservation in the Production of Petroleum.* New Haven, Conn.: Yale University Press, 1957.

U.S. Government Publications

Federal Energy Administration, *Project Independence Report.* Washington, D.C.: Government Printing Office, 1974.

Federal Trade Commission, *International Petroleum Cartel.* A Report to the Subcommittee on Monopoly. Select Committee on Small Business. 82nd Congress, 2nd Session, 1952.

Garland, T.M., Carrles M. Jr., and Convey, J.S., 'Assessment of U.S. Petroleum Supply with Varying Drilling Efforts', *Bureau of Mines Information Circular* 8634/1974.

Measday, Walter, 'A History of Federal Cooperation with the Petroleum

Industry', in U.S. Senate, Committee on Judiciary. Subcommittee on Antitrust and Monopoly, *Governmental Intervention in the Market Mechanism: The Petroleum Industry.* Washington, D.C.: Government Printing Office, 1969.

National Petroleum Council, *U.S. Energy Outlook.* Washington, D.C.: Government Printing Office, 1972.

U.S. Congress, House. Committee on Banking, Currency and Housing, *The Economics of Energy and Natural Resource Pricing.* Washington, D.C.: Government Printing Office, 1975.

U.S. Congress, House. Committee on Energy and Commerce, *The Changing Distribution of Industrial Profits: The Oil and Gas Industry Within the Fortune 500, 1978-80.* Washington, D.C.: Government Printing Office, 1981.

U.S. Congress, House. Committee on Science and Astronautics. Subcommittee on Energy, *Energy Facts.* Washington, D.C.: Government Printing Office, 1973.

U.S. Congress, House. Select Committee on Small Business, *The Third World Petroleum Congress.* 82nd Congress, Washington, D.C.: Government Printing Office, 1952.

U.S. Congress. Joint Committee on Atomic Energy, *Understanding the National Energy Dilemma.* Washington, D.C.: Government Printing Office, 1974.

U.S. Congress, Senate. Committee on Energy and Natural Resources, *Access to Oil—The United States Relationships with Saudi Arabia and Iran.* Washington, D.C.: Government Printing Office, 1977.

U.S. Congress, Senate. Committee on Finance, *Energy Statistics.* Washington, D.C.: Government Printing Office, 1975.

U.S. Congress, Senate. Committee on Foreign Relations, *Hearings, Multinational Corporations and United States Policy.* Washington, D.C.: Government Printing Office, 1974.

U.S. Congress, Senate. Committee on Government Operations, *Current Energy.* Shortage Oversight Series. Washington, D.C.: Government Printing Office, 1974.

U.S. Congress, Senate. Committee on Interior and Insular Affairs, *Estimate of the Economic Cost of Producing Crude Oil.* Washington, D.C.: Government Printing Office, 1976.

U.S. Congress, Seante. Committee on Interior and Insular Affairs, *Federal Leasing of Petroleum on the Outer Continental Shelf.* Washington, D.C.: Government Printing Office, 1976.

U.S. Congress, Senate. Committee on Interior and Insular Affairs, *Report of the Federal Trade Commission on Federal Energy Land Policy: Efficiency, Revenue, and Competition.* Washington, D.C.: Government Printing Office, 1976.

U.S. Congress, Senate. Committee on the Judiciary. Subcommittee on

Antitrust and Monopoly, *Economic Concentration.* 88th Congress. Washington, D.C.: Government Printing Office, 1964.

U.S. Congress, Senate. Subcommittee on Antitrust and Monopoly, *Hearings on Governmental Intervention in the Market Mechanism.* Washington, D.C.: Government Printing Office, 1969.

U.S. Congress, Senate. *U.S. Oil Companies and the Arab Oil Embargo: The International Allocation of Constrained Supplies.* Washington, D.C.: Government Printing Office, 1975.

U.S. Department of Commerce, *Survey of Current Business.* September, 1973.

U.S. Department of Energy, *An Analysis of the Productivity of Domestic Petroleum Exploration Activities.* Springfield, Virginia: National Technical Information Service, 1978.

U.S. Department of Energy, *1981 Annual Report to Congress,* vol. 2. Washington, D.C.: National Energy Information Center, 1982.

U.S. Department of Energy, *An Evaluation of Future World Oil Prices.* Springfield, Virginia: National Technical Information Service, 1978.

U.S. Department of Energy, *Depth and Producing Rate Classification of Domestic Oil Reservoirs, 1974.* Washington, D.C.: Government Printing Office, 1978.

U.S. Department of Energy. Division of Oil and Gas Analysis, *Factors Affecting the Decline Rate of Old Oil Production.* Washington, D.C.: National Technical Information Service, 1978.

U.S. Department of Energy, *Monthly Energy Review.* Washington, D.C.: Government Printing Office, various issues.

U.S. Department of Energy, *The Role of Foreign Governments in the Energy Industries.* Washington, D.C.: Office of International Affairs, 1977.

U.S. Department of the Interior. Bureau of Mines, *Mineral Yearbook.* Washington, D.C.: Government Printing Office, various years.

U.S. Department of the Interior, 'Depth and Producing Rate Classification of Oil Reservoirs in the 14 Principal Oil Producing States', *Bureau of Mines Information Circular* 8362. Washington, D.C.: Government Printing Office, 1967.

U.S. Department of the Interior, 'Depth and Producing Rate Classification of Petroleum Reservoirs in the United States, 1971', *Bureau of Mines Information Circular* 8675A. Washington, D.C.: Government Printing Office, 1976.

U.S. Department of the Interior, 'Land Utilization and Reclamation in the Mining Industry, 1930–71', *Bureau of Mines Information Circular* 8642. Washington, D.C.: Government Printing Office, 1974.

U.S. Department of the Interior, 'Offshore Petroleum Studies', *Bureau of Mines Information Circular* 8575. Washington, D.C.: Government Printing Office, 1973.

Journals and Newspapers

Adelman, M.A., 'Oil Prices in the Long Run, 1963-75', *Journal of Business,* April, 1964.
Adelman, M.A., 'Oil in the Eighties', *Petroleum Economist,* October, 1980, pp. 414-418.
Alnasrawi, Abbas, 'Collective Bargaining Power in OPEC', *Journal of World Trade Law,* March-April 1973, pp. 188-207.
American Petroleum Institute, 'Key Financial Data of Lending U.S. Oil Companies, 1968-78', *Discussion Paper* No. 017, September, 1979.
Anderson, Jack and Whitten, Les, 'Kissinger Allowed Oil Prices to Soar', *The Washington Post,* July 7, 1977, p. D15.
'Another Increase Expected in Exploratory Drilling', *World Oil,* February 15, 1975, pp. 84-85.
'Average World Wide Oil Finding Cost. . .', *Petroleum Outlook* 28, (May 1975): 36-38.
Barrows, Gordon H., 'Special Report on World Petroleum Concessions' *Petroleum Economist,* October 1980, pp. 426-428.
'Case for a "Rapid Deployment Force" ', *Petroleum Economist,* April 1981, pp. 134-35.
Chevalier, Jean-Marie, 'The Problem of Control in Large American Corporations', *Antitrust Bulletin,* Spring 1969.
Clifton, James, 'Competition and the Evolution of the Capitalist Mode of Production', *Cambridge Journal of Economics* 1 (June 1977): 137-151.
Copp, Anthony E., 'Technical Change and Petroleum Drilling Costs', *Land Economics* 50 (May 1974): 120-125.
DaBai, George S., 'Petroleum Finding Costs in United States', *Quarterly of the Colorado School of Mines* 70 (April 1975): 1-38.
DeNovo, John, 'The Movement for an Aggressive American Oil Policy Abroad, 1918-1920', *American Historical Review,* July 1956.
'Dependency Theory: A Reassessment', Special Issue, *Latin American Perspectives* 1 (Spring 1974).
'Dependency and Marxism', *Latin American Perspectives* 8 (Summer and Fall 1981).
'Energy Crisis in America', *Congressional Quarterly,* Winter 1973.
'Expanded Drilling Won't Solve the Energy Dilemma', *World Oil,* October 1978, pp. 133-138.
The Economist, various issues.
Fine, Ben, 'On Marx's Theory of Agricultural Rent', *Economy & Society* 8 (August 1979): 241-278.
Fitch, Bob, 'OPEC: The Big Cartel That Couldn't. . .', *Against the Current,* Spring 1982, pp. 16-22.
Flanigan, James, 'Acreage—the Newest Game in Oil', *Forbes,* September 1979, pp. 116-120.

Fliakos, C. and Lewison, R., 'Prospects for International Oil Supply and Demand: 1975, 1980, 1985', *The Journal of Energy and Development*, Autumn 1975, pp. 69–83.

Gault, John, 'Comparison of Petroleum Tax Systems', *Petroleum Economist*, October 1978, pp. 409–411.

Girvan, Norman, 'Economic Nationalism', *DAEDALUS* 104 (Fall 1975): 145–158.

Gordon, Richard L., 'A Reinterpretation of the Pure Theory of Exhaustion', *Journal of Political Economy* 75 (1967): 274–86.

'Government's Role in Oil Policy', *Petroleum Economist*, March 1981, pp. 90–93.

Gray, L.C., 'Rent Under the Assumption of Exhaustibility', *Quarterly Journal of Economics* 28 (1914): 466–489.

Greider, William and Smith, J.P., 'A Proposition: High Oil Prices Benefit U.S.', *The Washington Post*, July 10, 1977, p. 1.

'Higher Prices Seen Vital to Prevent Expected Crisis', *Oil and Gas Journal*, October 19, 1970, pp. 43–44.

'Hudson Institute Attacks British Attitudes to North Sea Oil Reserves', *Petroelum Times*, November 29, 1974, p. 15.

Kahn, Alfred E., 'The Depletion Allowance in the Context of Cartelization', *American Economic Review* 54 (June 1964).

Koven, Ronald, 'OPEC Deadlocked on Oil Pricing Plan', *The Washington Post*, September 17, 1980, p. 29.

Lenczowski, George, 'The Oil Producing Countries', *DAEDALUS* 104 (Fall 1975): 59–72.

Levy, Walter J., 'World Oil Corporation or International Chaos', *Foreign Affairs*, July 1974.

Martin, Douglas, 'Shell's Success in Oil Bidding', *New York Times*, May 26, 1981, D1.

McKie, James W., 'The United States', *DAEDALUS* 104 (Fall 1975): 73–90.

Miller, Edward, 'Some Implications of Land Ownership Pattern for Petroleum Policy', *Land Economics* 49 (November 1973): 414–423.

Minard, Lawrence, 'On the Spot', *Forbes*, January 1980, pp. 29–30.

Murray, Robin, 'Value and Theory of Rent', Part 1, *Capital & Class*, Autumn 1977, pp. 100–122.

Murray, Robin, 'Value and Theory of Rent', Part 2, *Capital & Class*, Spring 1978, pp. 11–33.

Nordhaus, William, 'The Allocation of Energy Resources', *Brookings Papers on Economic Activity*, no. 3, 1973.

Odell, Peter R., 'A Personal View of "Missing Oil" ', *Petroleum Economist*, January 1980, pp. 21–23.

'Oil Crisis In Perspective', Special issue, *DAEDALUS: Journal of the American Academy of Arts and Sciences*, Fall 1975.

'The Oil Crisis is Real This Time', Special Report, *Business Week,* July 1979, pp. 44-60.

'Oil in the Eighties: Tight Supply, Soaring Capital Outlays', Special Report, *Oil and Gas Journal,* November 12, 1979, pp. 163-169.

'OPEC Loses Control of Prices', *Petroleum Economist,* January 1980, pp. 2-3.

Pahlavi, M.R., 'Shah Fought Big Oil Companies To Boost Iran's Share of Wealth', *The Washington Star,* September 3, 1980, p. 1.

Parker, R.L., 'Observation of a Driller', *International Petroleum Times,* April 1, 1979, pp. 11-14.

Penrose, Edith, 'Government Partnership in the Major Concessions', *Middle East Economic Survey.* (August 30, 1968).

Penrose, Edith, 'The Development of Crisis', *DAEDALUS* 104 (Fall 1975): 39-57.

Petroleum Economist, various issues.

Petroleum Intelligence Weekly, various issues.

Petroleum Outlook, various issues.

'Policy for the Middle East', *Petroleum Economist,* May 1980, pp. 186-187.

Prodi, Romano and Clô, Alberto, 'Europe', *DAEDALUS* 104 (Fall 1975): 91-112.

Rahmer, B.A., 'Successes and Setbacks for Alternatives', *Petroleum Economist,* June 1982, pp. 241-243.

'Resumed Growth in Capital Outlays', *Petroleum Press Service,* February 1972, p. 60.

Roeber, Joe, 'What Determines Crude Oil Prices?', *Petroleum Economist,* May 1982, pp. 181-183.

'Second Thoughts', *Forbes,* June 15, 1973, p. 88.

Semmler, Willi, 'Competition, Monopoly and Differentials of Profit Rates: Theoretical Consideration and Empirical Evidence', *Review of Radical Political Economics,* Winter 1982, pp. 39-51.

Semmler, Willi, 'Competition, Monopoly, and Differential Profit Rates—A Reconsideration of the Classical and Marxian Theories', *Rivista Internazionale di Scienze Economiche e Commerciali,* Agosto 1982, pp. 737-760.

Shaikh, Anwar, 'Neo-Ricardian Economics: A Wealth of Algebra, A Poverty of Theory', *Review of Radical Political Economics* 14 (Summer 1982), pp. 67-83.

Shaikh, Anwar, 'Political Economy and Capitalism: Notes on Dobb's Theory of Crisis', *Cambridge Journal of Economics* 2 (1978): 233-251.

Sherrill, Robert, 'Energy Crisis: The Industry's Fright Campaign', *The Nation,* June 26, 1972.

'Spending More Money, But Finding Less Oil', *Petroleum Outlook* 29 (May 1976): 13-15.

Sraffa, Piero, 'The Laws of Returns under Competitive Conditions', *Economic Journal,* 36 (1926): pp. 535–550.

Steele, Henry, 'Public Policy Problems of the Domestic Crude Oil Industry', *American Economic Review* 54 (March 1964).

Stevens, William K., 'U.S. Oil Rush Begins to Slow', *New York Times,* March 29, 1982, p. D1.

Stigler, George J., 'Perfect Competition, Historically Contemplated', *Journal of Political Economy,* 65 (1957): pp. 1–17.

'Stop Oil Doubling Again', *The Economist,* November 19, 1980, pp. 13–14.

'Symposium on Exhaustible Resources', *Review of Economic Studies,* 1974.

'The Taming of the Wildcatter', *Forbes,* October 2, 1978, pp. 87–89.

Thanh, P.C., 'Growth and Optimal Growth with Natural Resources', February 1977, Department of Economics, The American University. (Mimeographed).

Tucker, E.S., 'Revenues and Balances', *Petroleum Economist,* July 1978, pp. 285–286.

Tucker, Robert W., 'Oil: The Issue of American Intervention', *Commentary,* January 1975, pp. 21–23.

'U.S. Drilling: The Trend is Up', *World Oil,* February 15, 1975, pp. 76–82.

'U.S. is the Most Costly Region to Find and Develop Oil and Gas', *Petroleum Outlook* 32 (May 1979): 1–4.

Vernon, Raymond, 'An Interpretation', *DAEDALUS* 104 (Fall 1975): 1–14.

'Views on Dependency', *Latin American Perspectives* 6 (Spring 1979).

Walden, Jerrold L., 'The International Petroleum Cartel in Iran—Private Power and Public Interest', *Journal of Public Law* 11 (Spring 1962): 3–60.

Weeks, John, 'Marx's Theory of Competition and the Theory of Imperialism', Department of Economics, The American University, November 1977. (Mimeographed).

West, Jim, 'Independent Drilling Despite Uncertainties', A Special Report, *Oil and Gas Journal,* October 1979, pp. 97–103.

'Who Drills U.S. Wells?', *World Oil,* February 15, 1980, p. 129.

'Widening OPEC Differentials', *Petroleum Economist,* February 1979, pp. 65–66.

Wilson, John W., 'Market Structure and Interfirm Integration in the Petroleum Industry', *Journal of Economic Issues,* June 1975.

Wood, P.W.J., 'There's Trillion Barrels of Oil Awaiting Discovery', *World Oil,* June 1979, pp. 141–148.

Yamani, A.Z., 'Participation Versus Nationalization—A Better Means to Survive', *Middle East Economic Survey* 12 (June 13, 20, 1969).